T0140308

Spatio-Temporal Graph Data Analytics

Venkata M.V. Gunturi • Shashi Shekhar

Spatio-Temporal Graph Data Analytics

 Springer

Venkata M.V. Gunturi
Dept of Computer Science and Engineering
Indian Institute of Technology – Ropar
Rupnagar, Punjab, India

Shashi Shekhar
Dept of Computer Science and Engineering
University of Minnesota
Minneapolis, Minnesota, USA

ISBN 978-3-319-88486-8   ISBN 978-3-319-67771-2   (eBook)
https://doi.org/10.1007/978-3-319-67771-2

Printed on acid-free paper

This Springer imprint is published by Springer Nature
The registered company is Springer International Publishing AG
The registered company address is: Gewerbestrasse 11, 6330 Cham, Switzerland

Preface

Given the increasing proliferation of location enabled sensors, we are able to collect a wide variety of data which has spatio-temporal semantics. Many of these datasets have network semantics as well. For instance, traffic sensors on urban road networks enable us to collect data on traffic signal delays and traffic congestion. Sample upcoming datasets from these sensors include temporally detailed (TD) roadmaps which provide typical travel speeds experienced on every road segment for thousands of departure-times in a typical week. Likewise, we have temporally detailed (TD) social networks that contain a temporal trace of social interactions among the individuals in the network over a time window. These datasets can have huge societal impact. For instance, a 2011 McKinsey Global Institute report estimates that location-aware data could save consumers hundreds of billions of dollars by helping vehicles avoid traffic congestion via next-generation routing services such as eco-routing.

However, data generated on spatio-temporal graphs (STGs) presents significant challenges for the current computer science state of the art. First, they sometimes violate the cost function decomposability assumption of current conceptual models for representing and querying STGs. Second, they may also violate the stationary-ranking (of candidate solutions) assumption of dynamic programming based techniques such as Dijkstra's shortest path algorithm. This monograph discusses potential solutions to both these problems.

We start this monograph with a gentle introduction to STGs in Chap. 1. This chapter discusses few potential application domains for STGs. In Chap. 2, a discussion of some fundamental concepts underlying data is presented. A clear understanding of these concepts is important to ensure semantic correctness of the query results. Chapter 3 presents some representational models and data structures for STGs. Additionally, this chapter also presents a solution to the challenge of *non-decomposability* mentioned in the previous paragraph. Following this, in Chap. 4, we discuss several algorithms for computing the fastest path for a single departure-time. In addition, we also present an algorithm to compute betweenness centrality on STGs. Chapters 5 and 6 discuss few advanced routing-related concepts relating to STGs. In these chapters, we introduce the idea of critical-time-point based approaches which address the second challenge of non-stationary ranking of candidate paths in STGs. In Chap. 7, we present a case study on social interaction data highlighting the need for considering temporal disaggregation in knowledge discovery. Chapter 8 introduces a new upcoming spatio-temporal data generated through the sensors embedded on an engine. The chapter also presents few open research questions on this dataset.

Ropar, India
September 2017

Venkata M.V. Gunturi

Acknowledgements

There are many people who have earned my gratitude for their contribution towards making this monograph a reality. First, I would like to thank my adviser, Prof. Shashi Shekhar, for his mentorship, support and guidance through my PhD. I am sincerely grateful for all the learning opportunities he gave me which not only helped me develop my skills as a research scholar, but also taught me how to be a good communicator. I thank all the professors who have guided me over the years in my research, some of whom also served on my review committee: Prof. Vipin Kumar, Prof. Ravi Janardan, Prof. William Northop, Prof. Arindam Banerjee and Prof. Henry Liu.

I extend my gratitude to my collaborators Prof. Kathleen Carley and Dr. Kenneth Joseph at the Carnegie Mellon University for their open mindedness and patience in exploring the applications of spatio-temporal graph data analytics in the domain of social network analysis. I would also like to thank Reem Ali and Andrew Kotz at the University of Minnesota (UMN) for patiently exploring the future research directions of spatio-temporal graphs in the domain of vehicle measurement data. I would like to extend my special thanks to Prof. Shekhar's spatial computing research group at UMN for their feedback over the years. And last (but not least), I would like to thank Springer for their advice and help in formatting of this monograph.

This work was supported by the Indian Institute of Technology, Ropar, India, DST SERB grant (grant number ECR/2016/001053), NSF grants (grant number NSF IIS-1320580, NSF No. 0940818, NSF IIS-1218168) and USDOD grant (grant number HM1582-08-1-0017). The content does not necessarily reflect the position or the policy of the government and no official endorsement should be inferred.

Contents

Chapter 1
Introduction

Spatio-temporal graphs can appear in a number of application domains such as transportation (e.g., for use in route-planning algorithms), social science (e.g., for modeling geospatial aspect of different social phenomena), urban planning and public safety (e.g., for studying road accidents), commerce (trades between countries), etc. Out of these domains, transportation is likely to be the biggest consumer of the techniques developed for spatio-temporal graphs. In this context, route-planning algorithms would be the most obvious application. In addition, spatio-temporal graphs can be used for answering several other transportation related questions as well, examples include, "Which lanes should be reversed during what times of the day?" or "what should be red-light and green light durations of this traffic signal during morning rush hours?" Moreover, the concept of transportation is not just limited to road networks, it include airways and sea routes as well. Apart from transportation, spatio-temporal graphs can also be very beneficial in applications relating to social science. One of the them being, to study the contact graphs to estimate the spread of an unit of information or a particular type of contagious infection. Following are few sample spatio-temporal graphs in the domains of transportation and social science.

Outline of the Chapter This chapter introduces two application domains relevant to spatio-temporal graphs, viz., urban road networks (Sect. 1.1) and social networks (Sect. 1.2). For both these application domains, we provide sample questions which can be answered through spatio-temporal graphs.

1.1 Urban Road Networks

Figure 1.1 illustrates a sample data that is available for modern day urban road networks. Here, road segments are shown using directed edges and road intersections are shown using nodes. Number written on the edges represent its mean travel-time

© Springer International Publishing AG 2017
V.M.V. Gunturi, S. Shekhar, *Spatio-Temporal Graph Data Analytics*,
https://doi.org/10.1007/978-3-319-67771-2_1

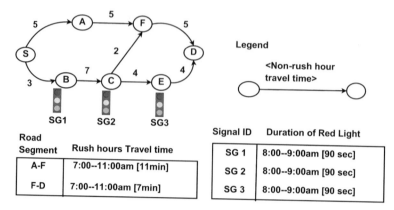

Road Segment	Rush hours Travel time
A-F	7:00--11:00am [11min]
F-D	7:00--11:00am [7min]

Signal ID	Duration of Red Light
SG 1	8:00--9:00am [90 sec]
SG 2	8:00--9:00am [90 sec]
SG 3	8:00--9:00am [90 sec]

Fig. 1.1 Sample urban road network (best in color)

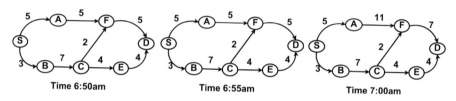

Fig. 1.2 Urban road network represented as a series of snapshots

(in minutes) during the non-rush hour. The figure also shows three traffic signals SG1, SG2 and SG3, which control the traffic moving along S-B-C-E-D. These signals have a red light duration of 90 s during 8:00am–9:00pm. Further, the mean rush hour (which starts at 8:00am) travel-time on segments A-F (11 min) and F-D (7 min) are also illustrated in the figure. In order to execute different kinds of query algorithms on this, we would have to represent this information in a form which can be easily consumed by a query processing algorithm. Figure 1.2 is one such representation (more covered in Chap. 3) of the underlying spatio-temporal graph. Here, the road network is represented as a series of snapshots at different times. Each snapshot represents the state of the network (i.e., the travel-times on its edges) at a particular time point. For sake of simplicity, Fig. 1.2 does not show the traffic signal information. Traffic signal delays can also included into a spatio-temporal graph. However, those models are little more involved and are covered in Chap. 3.

In general, it is important to make a note of the temporal granularity in representations such as the one shown in Fig. 1.2. For instance, one may note that in Fig. 1.2 the snapshots are at an interval of 5 min. This would raise challenges if the travel-time on road segments is not a multiple of 5, which would be the case in any realistic scenario. Having snapshots at 1 min time interval would elevate the issue. But, in case the original travel-time data is not available at a granularity of 1 min, then we may choose to replicate the data using a step function. For instance, consider a case where the available data is at a granularity of 15 min, i.e., we have travel-times on the edges at times 9:00am, 9:15am, etc. In the suggested scheme, the mean travel-time observed at 9:00am would be assumed to prevail during the entire time duration of 9:01–9:14am.

Sample Problems of Interest on Urban Road Networks Urban road networks are a key part in our daily lives. One of the most popular use-cases on road networks involve transportation. Following are few sample queries related to transportation that one may pose on road networks.

1. What is the fastest path to airport from the University for a 9:00am departure?
2. When is the latest I should depart from my home in order to reach the airport by 5:00pm?
3. What is the best time to depart so that I spend least amount of time traffic?

1.2 Social Networks

Figure 1.3 illustrates the location history of five people across timestamps $t = Jan$-1, $t = Jan$-2, $t = Jan$-3 and $t = Jan$-4. In this example, person B acquires a contagious infection (e.g., Ebola) when he was in location $loc1$ on Jan-1. After that he moves to location $loc2$ (and back to $loc1$) while still being infected. He gets cured by Jan-4, when he was at $loc4$. Combining this with the location information of other people, one can study the spread of the contagion across a set of people. This can be done by modeling the data as a spatio-temporal graph. Figure 1.4 illustrates the spatio-temporal graph constructed for this example. Here, each snapshot represents a collection of contacts observed on a day. Inside each snapshot, two nodes are connected by a link if they have been in contact. For instance, in our example, person C was in the same location as person E on Jan-1. This is shown as a undirected edge in the figure between C and E in the snapshot corresponding to Jan-1. While

	Jan 01	Jan 02	Jan 03	Jan 04	Jan 05
Node A	Location 3	Location 3	Location 3	Location 3*	Location 1*
Node B	Location 1*	Location 2*	Location 1*	Location 4	Location 4
Node C	Location 3	Location 3	Location 3	Location 4	Location 4
Node D	Location 2	Location 2*	Location 2*	Location 3*	Location 4
Node E	Location 3	Location 3	Location 3	Location 3*	Location 2*

Fig. 1.3 Location history of actors for a social contact graph

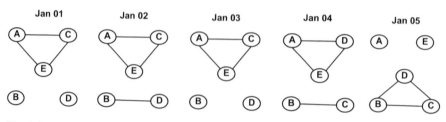

Fig. 1.4 A social network represented as series of snapshots

studying the spread of a contagion, we would pay special attention to these edges (and their time-coordinate) and evaluate them for their potential to contribute to the spread.

It is important to note that in case of social networks, depending on the application under consideration, one may treat each "snapshot" as a *panel* which is an aggregation of all the links over a certain period of time. For instance, in the example shown in Fig. 1.4, each snapshot was indeed an panel created by aggregating all the contacts that happened over a day. Window length of the panel would depend on factors like temporal granularity of the available data and nature of phenomenon being studied, etc. For instance, if we were to study the local social influence of group leader through re-tweets, we may limit the window length to few weeks or a month as his or her influence may not change much in that time (unless there are outliers). However, if we are tracking a potential terrorist or a spy, then we may be interested in his or her day to day contacts as well. In general, the window length of the panel can play a very important role in determining the usefulness of the obtained analysis. If the window length is too long, then it might obfuscate the "signal of interest". Similarly, if it is too narrow, then we may not catch it at all. More of this concept is covered in Chap. 7.

Similar to generalizing the concept of snapshots through the notion of panels, one can also generalize the *space* aspect of a social network by interpreting it in more abstract ways (depending on the needs of the application). For example, if one is interested in studying the local influence of actors in organizational setting, then departments may be a proxy for location.

Sample Problems of Interest on social Networks A spatio-temporal based analysis of social networks may help us gain valuable understanding of the underlying social phenomena. Following are few sample problems of interest. This monograph would cover few algorithms for these problems.

1. Are there any transient communities in the network?
2. Whose social influence (as measured through centrality) fluctuated the most during the Jan–July period?
3. Who all may need to be observed for symptoms of the infection?
4. Who are the key liaisons for information flowing from sales department to finance department?

Chapter 2
Fundamental Concepts for Spatio-Temporal Graphs

Data generated on a spatio-temporal graph implicitly captures several concepts related to the domain from where the data is originating. It is extremely important to acknowledge the semantics of the data while devising algorithms on spatio-temporal data. Failure to do so may lead to a unreliable results. For instance, while working with the travel-time data for a routing application, it is very important to consider, what is referred to as, Lagrangian reference frame (more details in Sect. 2.1). Similar is the case with social interaction data, if one is working with data relating to information flow, then one should work in Lagrangian reference frame.

Outline of the Chapter Section 2.1 presents some fundamental concepts pertinent to the urban road network datasets. In this section, we discuss concepts like reference fame and nature of property (*holistic vs decomposable*) recorded. We also discuss the duality between the data collection reference frame and data query frame. In Sect. 2.2, we present the fundamental concepts pertinent to social interaction data. Here, we discuss the ideas of panelization along the temporal dimension and reference frame in context of social interaction data.

2.1 Road Network Datasets

Urban road networks are embedded in space. Datasets on these networks are created by measuring some parameters on this physical entity. Consider the sample urban road network shown in Fig. 2.1 (on the left). Here, the arrows represent road segments and labels (in circles) represent an intersection between two road segments. Location of traffic signals are also annotated in the figure. On this network, we consider three types of datasets: (a) temporally detailed (TD) roadmaps [60], (b) traffic signal data [48], and (c) annotated GPS traces [60, 68]. These datasets record either historical or evolving aspects of certain travel related phenomena on our transportation network.

© Springer International Publishing AG 2017
V.M.V. Gunturi, S. Shekhar, *Spatio-Temporal Graph Data Analytics*,
https://doi.org/10.1007/978-3-319-67771-2_2

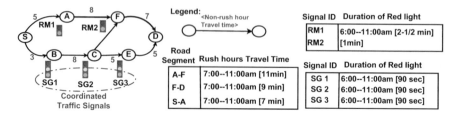

Fig. 2.1 Sample TD roadmap and traffic signal data

TD roadmaps store historical travel-time on road segments for several departure-times in a typical week [60]. For simplicity, TD roadmap data is illustrated in Fig. 2.1 by highlighting the morning rush hour (7:00am–11:00am) travel-time only on segments A-F, F-D and S-A. The number written on the edges represent the travel-time at other times of the day. The travel-times of other road segments (S-B, B-C, C-E, E-D, C-F) in the figure are assumed to remain constant throughout the day. The figure also shows the red-light duration of SG1, SG2, SG3, RM1 and RM2 traffic signals during the morning rush hour period. In our example, the traffic signals SG1, SG2 and SG3 are *coordinated for a journey from S to D*. This means that the red-light durations and phase gap among traffic signals SG1, SG2 and SG3 are set such that a typical traveler starting at S and going towards D (within certain speed-limits) would typically *wait only at SG1*, before being smoothly transferred through intersections C and E *without any waiting at SG2 or SG3*.

In addition to TD roadmaps and traffic signal data, one can also consider map-matched and pre-processed [69] GPS traces. These may also annotated with data from engine computers to get richer datasets illustrating fuel economy of the route. We refer to them as annotated GPS traces. Each trace in this dataset is represented as a sequence of road-segments traversed in the journey along with its corresponding schedule denoting the exact time when the traversal of a particular segment began. GPS traces can potentially capture the evolving aspect of our system. For instance, if segment E-D (Fig. 2.1) is congested due to an event (a non-equilibrium phenomenon), then travel-time denoted by TD roadmaps may no longer be accurate. In such a case, a traveler may prefer to follow another route (say C-F-D) which other commuters may be taking to reach D.

2.1.1   Key Concepts in Road Network Datasets

Road network datasets capture routing-related concepts along two dimensions: (i) frame of reference, (ii) nature of property recorded (see Fig. 2.2).

Frame of Reference For the first dimension, datasets are assembled with either *Eulerian* or *Lagrangian* reference frame shown upfront. In Eulerian reference frame, the phenomenon is observed through fixed locations in the system [4]. For

Fig. 2.2 Taxonomy of
concepts

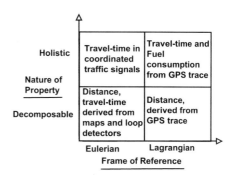

a transportation network, this view corresponds to a stationary observer sitting on a side of a freeway (loop detectors or traffic observatory). By contrast, Lagrangian frame of reference corresponds to the perspective of an traveler driving along a particular route [4]. Data collected through annotated GPS traces is Lagrangian in nature. Both Eulerian and Lagrangian reference frames can be thought of as means to represent a set of unique space-time coordinates in the transportation network. Generally, we can consider all records in any road network dataset as certain attributes associated with these space-time coordinates. For example, a data record containing the measurement 5 min as travel-time on the road S-A at 9:00am on a Monday (TD roadmaps) of a typical week has the physical location of S-A for space and 9:00am-Monday for time coordinate respectively. Travel-time in TD-roadmaps shows Eulerian frame upfront, whereas in a annotated GPS trace it shows Lagrangian frame upfront.

Nature of Property   Under Eulerian or Lagrangian frame of reference, the property being recorded can be either *decomposable* or *holistic*. A decomposable property when re-constructed for a larger instance (e.g. route) by joining values for smaller instances (at appropriate space-time coordinates) retains its correctness. Distance measured from a GPS trace (Lagrangian frame) or as measured through maps (Eulerian frame) is a decomposable property. Other examples include, travel-time obtained from loop detectors (TD roadmaps), signal delays (red light duration) at individual traffic signals as set by traffic managers.

Decomposable properties can be further classified as deterministic or non-deterministic. Properties which can be expressed clearly from a start-point are termed as deterministic. Travel-time as captured in TD roadmaps is deterministic as given a departure-time (in a typical week), TD roadmaps will provide a unique travel-time for a road-segment. In contrast, traffic signal delays are non-deterministic unless the absolute start-time of their cycles is known. In other words, given a departure-time on a route, we cannot determine (at least in current datasets) the precise delay to be experienced at a signal.

On the other hand, holistic properties are properties when measured over a large instance cannot be meaning fully broken down into corresponding values for smaller instances through space-time decomposition. Consider again the traffic signal coordination scenario amongst signals SG1, SG2 and SG3 (in Fig. 2.1) which control the incoming traffic from segments S-B, B-C, and C-E (and going towards D) respectively. As mentioned earlier, the red-light durations and phase gap among traffic signals SG1, SG2 and SG3 are set such that a typical traveler starting at S and going towards D (within certain speed-limits) would typically *wait only at SG1*, before being smoothly transferred through intersections C and E *with no waiting at SG2 or SG3*. This means that, in this journey, an initial waiting at SG1 renders SG2 and SG3 wait free. Consider a typical journey from S to D through SG1, SG2 and SG3. If we were to decompose this journey into its component journeys S-B, experience at SG1, B-C, experience at SG2, C-E, experience at SG3 and C-D, then would not see any waiting at SG2 and SG3. This is, however, *not true* as a typical traveler starting intersection C (or E) would typically wait for some time SG2 (or SG3). We refer to this kind of behavior, where properties (e.g. travel-time) measured over larger instances (e.g. a route) loose their semantic meaning on decomposition, as *holism*. For our previous example, we say that the total travel-time measured over the route S-B-C-E-D was behaving like a *holistic property*. Apart from coordinated traffic signals, travel-time (and fuel consumption) experienced by a commuter on a road-segment, as derived from his/her annotated GPS trace, is holistic in nature. Here, the travel-time experienced would depend on his/her initial velocity gained *before* entering the particular road-segment under consideration. Travel-time of the commuter is deterministic as we can exactly measure its value from start of journey. By contrary, travel-time through a series of coordinated traffic signals is non-deterministic.

2.1.2   Data-Collection vs Querying Reference Frame in Road Networks

'Data-collection' reference frame may be different from 'querying' reference frame. For instance, travel-time recorded in TD roadmaps shows Eulerian frame upfront. Whereas, routing queries on the same TD roadmaps (and other road network datasets) would more meaningful through Lagrangian frame of reference [13, 27]. For instance, consider a case where we want to determine the typical travel time on route A-F-D for 6:52am departure in the network shown in Fig. 2.1. In a Eulerian frame, this would be sum of the travel times of individual segments, A-D (8 min), F-D (7 min), waiting time (1 min) at ramp meter RM2, giving a the typical travel-time experience to be between 15 and 16 min. However, this result is not correct, as once the traveler reaches segment F-D, its 7:00am and the morning travel time of the segment should be considered (9 min, refer Fig. 2.1). Thus, in this scenario we have to evaluate the route through a Lagrangian frame of reference. This frame would give the typical experience would be between 17 and 18 min. The interoperability of Eulerian and Lagrangian frame of reference allows for the 'querying' and 'data-collection' reference frames to be different. In other words, data recorded in one frame can be easily queried from (or converted to) another reference frame [4].

2.2 Social Network Datasets

Unlike road networks which are embedded in a physical space, social networks can at best be assumed to be present in a virtual space which is guided by some sociological laws. As a result, physical measurements on these networks is not possible. However, given the ever-increasing proliferation of mobile and location enabled social networking platforms, it has now become possible to measure many more parameters on these networks (in comparison to traditional questionnaire based data collection techniques) in an indirect way by recording the social interactions seen on these platforms. Social interactions such as emails, friend requests, post-comment interactions, co-location pairs, etc.

A social interaction can be described as a tuple $<A_i,A_j,T,loc>$, where A_i and A_j are the two individuals (or agents) who were interacting, and T is the time-stamp when this was observed. Here, both A_i and A_j could have an intrinsic space location associated with them. This, based on the needs of the application, can either be a physical location such as city, country etc., or a virtual location such as departments, organizations. In addition, the interaction can also have a *loc* attribute with the interaction which denotes the location where the particular social interaction occurred.

2.2.1 Key Concepts in Social Network Datasets

Any social interaction can be assumed to create a temporary link between the participating nodes (say A_i and A_j) with some space-time context, which can be analyzed to derive some insights. However, one must pay attention to some of the following concepts to ensure meaningful and reliable results.

Panelization Along Temporal Dimension We can either consider the social interactions individually (e.g. change detection in streams) or could first create a set of panels containing interactions among members of a specific social system (e.g., employees of a university). Here, each panel would either contain interactions that occurred at a specific time point or comprise of all the interactions spread over a certain time duration (e.g., hour, day, week, etc.). A collection of all such panels (over a time window) recording interactions among the members of the same social system is referred to as a *Temporally Detailed (TD) social network* in this monograph. Figure 2.3 illustrates a sample TD social network among W, X, A, C, D, U and Y. Here, an edge between two individuals denotes a social interaction.

Reference Frame Unlike the road networks where the (same) travel-time data can be collected through both Eulerian and Lagrangian reference frame, social network data collection is mostly done through Eulerian frame of reference; scientists typically download it from a central server. However, similar to road network datasets, query reference frame is an important aspect while studying social network datasets.

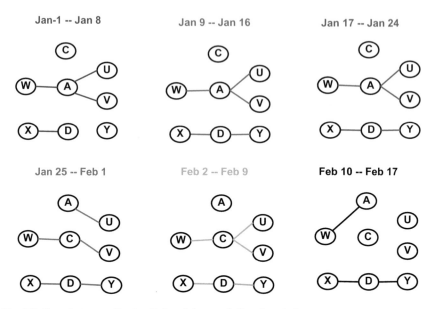

Fig. 2.3 Sample temporally-detailed social network (best in color)

In cases where we are interested in studying the flow of information in a network, we need to consider the temporal ordering among the social interactions. In other words, we need to consider Lagrangian reference frame while studying information flow. This aspect was highlighted by many works [31, 44, 47, 56, 61] which focused on studying information flow in a social network. Consider the top right panel in Fig. 2.3. As a contrived example, assume that all the interactions between node A and node U happened between dates Jan 17–Jan 19. Similarly, assume that all the interactions between node W and A happened during the time period Jan 20–Jan 24. Now, though the panel shows the path $W \rightarrow A \rightarrow U$ as a potential path for information flow, in reality it is not the case. This of course raises the question of ideal window length of the panel. Unfortunately, there is no perfect answer to that question; it depends on the question being explored in the dataset. A wide panel is likely to overestimate the number of paths greatly. Considering temporal order among social interactions is also not as straight forward. It raises the issue of considering waiting in the candidate paths. If node X sent an email to node Y on Jan-1, and node Y sent an email to node Z on Jan-4 (on the same or a similar subject), would both these events be considered as part of the same information flow chain? How much gap is acceptable? It is important to note that issue of waiting would not happen in case of re-tweets.

On the other hand, if the goal is to study the community structure (and its dynamics) created out of social interactions, then temporal ordering is not of much importance. In other words, the Eulerian frame of reference is most suited for studying some of the structural properties such as temporally evolving patterns of activity of individuals [7], creation of temporally linked structures [17], addition (and activity patterns) of social links [44] etc. Here, each panel can be composed of all the social interactions which happened over a certain period of time.

2.3   Conclusion

To summarize, spatio-temporal graph datasets capture a wide variety of concepts in them. In case of urban road network datasets, one must consider if the data can be decomposed or if it needs to be considered holistically. Also, depending on the query, one needs to choose between Eulerian and Lagrangian reference frame. Similarly, social interaction data also has its own unique aspects to be considered. Depending on the nature of the question being investigated, we need to consider aspects like length of the window to create panels and a notion of waiting while considering information flow (in Lagrangian reference frame) over the underlying social network.

Chapter 3
Representational Models for Spatio-Temporal Graphs

Designing suitable representational models for spatio-temporal graphs is a challenging task. On one end, the model should be expressive enough to be able to easily represent a wide variety of concepts (e.g., holistic properties and Lagrangian reference frame) captured in the underlying datasets on urban road networks and social networks. But on the other end, the model should also aid in developing computationally scalable algorithms.

Outline of the Chapter In this chapter, we would first describe the representational models at a more conceptual level (Sect. 3.1) which capture the required concepts in an easy to understand manner (from a visual sense), but are not computationally scalable. In this section, we describe both traditional models such as the time-expanded graphs, and the recently proposed Lagrangian Xgraphs (Sect. 3.1.1). Following this, we discuss models (Sect. 3.2) which more conducive for computationally scalable algorithms. In this section, we present a data structure for spatio-temporal digraphs.

3.1 Models at Conceptual Level

Consider a space which is a Cartesian product between set of physical entities and time coordinates. Each ordered pair in this Cartesian product –consisting of a unique physical entity and a unique time coordinate –would form a point in this space. For instance, consider A, B, C, etc., as entities of interest (road intersections or people in an organization), and these are observed over times t_1, t_2, etc. Then, the space created out of the Cartesian product would contain one point for each of the following pairs, (A at t_1), (A at t_2), (B at t_1), (B at t_2), etc. Conceptually, any dataset (generated either on a road network or a social network) which has spatio-temporal graph semantics can be represented in this space. These are detailed in this section.

© Springer International Publishing AG 2017
V.M.V. Gunturi, S. Shekhar, *Spatio-Temporal Graph Data Analytics*,
https://doi.org/10.1007/978-3-319-67771-2_3

Given this space of Cartesian product, we have following two kinds of models: (a) Eulerian-view models and (b) Lagrangian-view models. As mentioned in Chap. 2, Eulerian-view corresponds to the view of a person sitting in a fixed position and observing the network (or a portion of it) over time. With the intention of carrying this forward into our models, in the Eulerian-view models, relations (edges) are allowed only between points which either belong to the same physical entity (seeing the entity over time), or have the same time-coordinate (snapshot of the network at one time point). Lagrangian-view does not have this restriction on relations among points of the Cartesian product; a moving observer can view the network while traveling through space and time.

Eulerian-view models help put up the Eulerian reference frame upfront. Thus, these are most useful for answer snapshot based question, for e.g., "Who all have contracted the infection in the Jan-1–Jan-3 time period?" and "which road segments have a travel time of greater than 20 min at 9:00am?" Note that for answering the second question in a Eulerian-view model, Cartesian product should be between the set of road-segments and the set of departure-times. Figures 1.2 and 1.4 in Chap. 1 can be considered as Eulerian-view models in a slightly informal sense. Figure 3.1 shows a Eulerian-view of Fig. 1.2 in a slightly more formal sense. Here, a node represents the particular road-segment for the particular time coordinate. Travel-time information is associated with the nodes. Edges can be associated with turn delay (or red-light duration) information. For sake of simplicity, some nodes are not shown in this figure. Figure 3.2 illustrates a sample social contact graph among individuals W, A, U, V and D using a Eulerian-view model.

In contrast to Eulerian-view models, Lagrangian-view models put the Lagrangian view upfront. These have been quite popular in both the road network [39, 43] and the social network [30, 40] literature. Though the literature has several variants of Lagrangian-view models, we would describe only a basic version to keep the discussion simple. This basic version, referred to as *time-expanded graphs*, can represent only deterministic (e.g., travel time in temporally-detailed roadmaps)

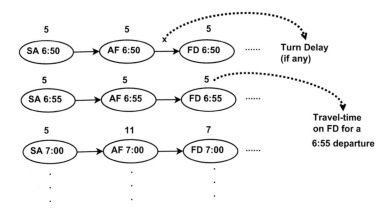

Fig. 3.1 Eulerian-view model representing travel-time in an urban road networks

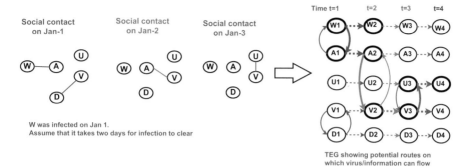

Fig. 3.2 Eulerian-view model representing spatio-temporal social contacts

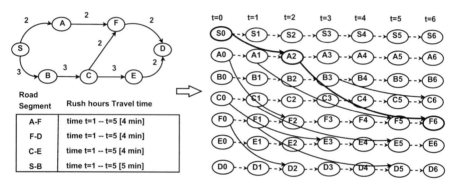

Fig. 3.3 Lagrangian-view model (time-expanded graph) representing travel-time in an urban road networks. Some edges are not shown to retain the readability of the figure

properties. A more generalized version of time-expanded graphs, referred to as *Lagrangian Xgraphs* [25] is covered later which is capable of representing non-deterministic properties as well.

Time-Expanded Graphs Consider the sample road network dataset shown in Fig. 3.3 (on the left). The figure mentions the rush-hour (departure-times in [1 5]) travel-time on road segments A-F, F-D, C-E and S-B. The non-rush hour travel-times are shown on the edges in the top left representation of the figure. A time-expanded graph (TEG) represents this data in the space of Cartesian product between road-intersections and departure-times. Figure 3.3 (on the right) illustrates this space. The travel-time information is represented in a TEG by adding edges between the appropriate nodes in the Cartesian product. For instance, a rush-hour travel-time of 4 min on the edge A-F, for a departure-time of $t = 1$, is represented by adding an edge between the nodes A1 and F5. It should be noted any path in a time-expanded graph gives a Lagrangian view. For instance, consider the path S-A-F for a departure-time of $t = 0$. This path is represented through the edges $(S0, A2)$ (non-rush hour travel-time of 2 min) and $(A2, F6)$ (rush-hour travel-time of 4 min) in TEG (shown in bold). Notice that this path automatically considers the rush-hour travel-time on edge A-F as rush hour has already begun by the time the traveler arrives on node *A*. TEGs can also used for social network datasets. They are particularly preferred for datasets where we are recording a directed flow of information, e.g., emails and tweets.

3.1.1 Lagrangian Xgraphs

The primary limitation of time-expanded graphs spawns from its binary nature of edges. This limits its capability to model non-deterministic properties such as signal delay observed at a traffic signal, which is intrinsically an n-ary relation among the nodes in the space resulting from a Cartesian product between spatial locations and time coordinates. In a nutshell, Lagrangian Xgraphs are a generalized version of time-expanded graphs where edges involve more than two nodes in the space of Cartesian product. Following is a formal definition of Lagrangian Xgraphs (replicated from [25]).

Given a space of Cartesian product between the spatial locations and time-coordinates, a Lagrangian Xgraph ($\mathcal{L}a\mathcal{X}$) as an ordered set of Xnodes ($\mathcal{X}v$), Xedges ($\mathcal{X}e$) and a time horizon parameter $\mathcal{H}$, i.e., $\mathcal{L}a\mathcal{X} = (\mathcal{X}v, \mathcal{X}e, \mathcal{H})$.

Xnodes Xnodes $\mathcal{X}v_i \in \mathcal{X}v$ represent the ordered pairs in the resultant Cartesian product between spatial locations and time-coordinate. For a road network dataset, each $\mathcal{X}v_i$ would be a particular road segment (spatial location) for a specific departure-time (time coordinate). Additionally, it would also be associated with other auxiliary information according to application needs.

Xedges are collection of Xnodes which are used to express a 'Lagrangian' relationship among a set of Xnodes. More formally, $\mathcal{X}e^i = \{\mathcal{X}v_s, \mathcal{X}v_1, \ldots, \mathcal{X}v_k, \mathcal{X}v_{d1}, \mathcal{X}v_{d2}, \ldots, \mathcal{X}v_{dj}\}$. Here, the first Xnode ($\mathcal{X}v_s$) and the last set of Xnodes ($\mathcal{X}v_{d[1\ldots j]}$) correspond to the instance of the first and the last entities participating in the relation and are marked separately for ease.

Fig. 3.4 Taxonomy of Xedges

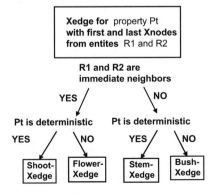

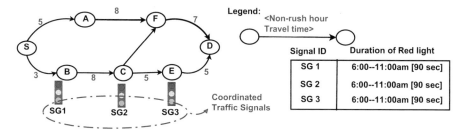

Fig. 3.5 Sample coordinated traffic signal data

Taxonomy of Xedges Xedges as defined above are further categorized according to:
(a) spatial relationship between the first ($\mathcal{X}v_s$) and last set of Xnodes ($\mathcal{X}v_{d[1\ j]}$) and
(b) property being modeled is deterministic or non-deterministic. Figure 3.4 shows
the proposed classification. Primarily, we classify an Xedge as: (i) shoot-Xedge,
(ii) flower-Xedge, (iii) stem-Xedge or, (iv) bush-Xedge. The first and last Xnodes in
both shoot-Xedges and flower-Xedges belong to spatial locations which are spatially
immediate neighbors. They would be used for modeling decomposable properties.
In a shoot-Xedge, each spatial location (e.g., a road segment or an intersection) is
present for a single time and thus, can model deterministic properties. On the other
hand, Flower-Xedges allow the same spatial location to be represented for multiple
time coordinates and thus, can be used to model non-deterministic properties (e.g.
individual traffic signal delays).

By contrast, both stem-Xedges and bush-Xedges allow the first and last
Xnode(s) to be separated physically. This physical separation allows both stem
and bush-Xedges to represent holistic properties. Bush-Xedges are useful for non-
deterministic properties since they allow one spatial location to be represented at
multiple time coordinates. Stem-Xedges are employed for deterministic properties
since each entity is present only for one time coordinate.

Lagrangian Xgraph for Road Network Datasets We now describe an instance of
Lagrangian Xgraphs modeling road network datasets (e.g., TD roadmaps and traffic
signal delay information). Here, Xnodes represent the road segments at multiple
departure-times and Xedges express the experience of a traveler among a sequence
of Xnodes. Here, each Xnode is associated with information such as: (a) start
and end point of the road-segment, (b) departure-time, (c) typical travel-time on
segment for the departure-time (from TD roadmaps). With the intention of keeping
the discussion focused, we only illustrate Xedges for a non-deterministic holistic
property, e.g., travel-time experienced in a sequence of coordinated traffic signals.

Coordinated Traffic Signals Consider the example of coordinated signals SG1,
SG2 and SG3 in Fig. 3.5. Here, we will represent typical experiences of journeys
through these signals starting from S at 7:00:00am. Given that signal delays are
non-deterministic and travel-time experienced in a sequence of coordinated traffic
signals is holistic, we would use bush-Xedges. Figure 3.6 illustrates bush-Xedges
for some journeys at 30 s temporal granularity. We have bush-Xedges for following

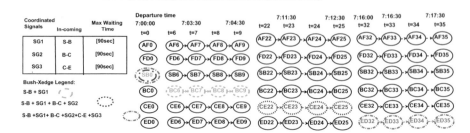

Fig. 3.6 First and last Xnodes for journey through a series of coordinated signals

journeys: (a) start at S and travel to the beginning of segment B-C (after SG1), (b) start at S and travel to the beginning of segment C-E (after SG2), (c) start at S and travel to the beginning of segment E-D (after SG3).

For case (a), the bush-Xedge would include (SB0) and (BC6, BC7, BC8 and, BC9) as Xnodes for the first $(\mathcal{X}v_s)$ and last Xnodes $(\mathcal{X}v_{d[1\ j]})$. Intuitively, this Xedge means that a typical journey starting from S at 7:00am along S-B (SB0 in the figure) can start traversing road segment B-C at times 7:03:00 (no wait at SG1, BC6 in figure), 7:03:30 (30 s wait at SG1, BC7 in the figure), 7:04:00 or 7:04:30 (90 s wait at SG1, BC9 in the figure). Similarly, in case (b), we would have SB0 and (CE22, CE23, CE24, CE25) as the first and last Xnodes. Internally, this would include the Xnodes (BC6, BC7, BC8 and, BC9) (not shown in figure to maintain clarity). Other bush-Xedges can be defined in a analogous way. Similarly, there would be other bush-Xedges representing journeys starting at beginning of B-C (after SG1).

Discussion Lagrangian Xgraphs have a flavor of hypergraphs but they differ from both general hypergraphs [5], which represent subsets of nodes without any reference frame and directed hypergraphs [20] which directly connect a set of sources to a set of destinations.

In theory, similar to Lagrangian Xgraphs, one can also define a concept of Eulerian Xgraphs. However, it was avoided as an Eulerian Xgraph would not have been of much use for the application domain questions considered in this monograph. At best, it could have been used for modeling group interactions in social network dataset, but any group interaction can also be represented by decomposing it into clique of the same size. Similarly, an email sent to many people can also be decomposed into individual directed edges in a TEG between the source and the recipients.

3.2 Data Structures for Algorithms

Eulerian-view and Lagrangian-view models discussed in the previous section are great to understand road network and social network datasets from a conceptual perspective. This is essential to ensure interpretability of the results. For instance, an overestimation on the number of likely information paths, caused when we ignore the Lagrangian reference frame in email data, would inflate the centrality of certain individuals in the network. However, the models discussed in the previous section, in their current form, would not immediately lead computationally efficient algorithms. The primary reason for this being, replication of the information of the physical-entity across the time coordinates. For example, in Fig. 3.3 (and Fig. 3.6), the road intersections (and road segments in Fig. 3.6) S, A, B, etc., were replicated across 6 departure-times. Such replication would lead to significant computational bottlenecks in any realistic road network dataset containing typical travel-time information for thousands of departure-times (in a typical week) over almost all the road segments in the city.

For this reason, many works [9–11, 14, 16, 16, 22, 23, 30, 34, 57, 64] "compressed" these conceptual models into data structures which were space efficient and could support algorithms that easily scaled-up to city scale datasets. In all these data structures, the basic idea was to work with the underlying network (road or social), and then to associate functions with each edge that could capture the time-varying nature of its properties (e.g., travel-time); typically one function for each property. Also, if needed, nodes could also have functions for representing its time-varying properties. Depending on the data available, some works [11, 14, 16, 34] opted for continuous functions, whereas others [9, 10, 22, 23, 30, 64] preferred discrete functions. We now describe a basic version of these data structures graphs which we refer to as *temporal digraphs*. Like the other data structures proposed in the literature, this can also only model deterministic properties. Following this, we discuss a slightly more complicated representation which can model both deterministic and non-deterministic properties. In this, we also illustrate holistic properties.

3.2.1 Temporal Digraphs

Figure 3.7 illustrates the said temporal binary-edged digraph for a road network. Here, nodes represent the road intersections and edge represent the road segments between two road intersections. The time series represented on edges denote the

Fig. 3.7 A temporal binary-edges digraph representing travel-time information in a road network

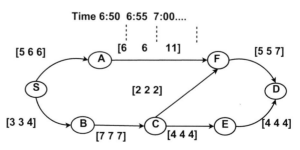
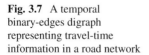

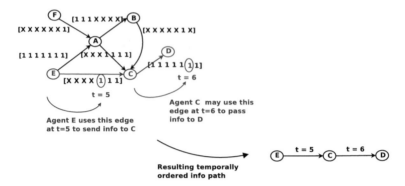

Fig. 3.8 A temporal binary-edged digraph representing email data in a social network

travel-time for departure-times. For instance, the time series [5 5 11] shown on edge A-F represents the fact that travel-time on this edge is 6, 6 and 11 time-units for a departure of 6:50am, 6:55am and 7:00am at node A. In principle, this is a temporally-detailed roadmap discussed in Chap. 2. This travel-time information on a edge can be easily implemented in any high-level programming language using constructs like hash maps. And in turn all the hash maps (one for each edge) can be put in another hash map.

While the described model seems intuitive, one must also consider the following issue before using this data structure in any algorithm. Many times the available dataset would have travel-time information only at low temporal granularity, for e.g., at a 5 min resolution. In such cases, creating a route and determining its total-travel time would be challenging. For instance in Fig. 3.7, if one departs at node A at 6:50am, he would reach node F at 6:56am. However, travel-time information on edge F-D is available only for 6:50am, 6:55am and 7:00am. In order to avoid such situations, one should interpolate the travel-time for intermediate departure-times using either a step function or a linear interpolation [14].

Figure 3.8 illustrates email data using a temporal binary-edged digraph. Here, nodes correspond to individuals in a social network, and directed edges represent social connections. A time-series is associated with each edge which stores information on emails sent along that particular social connection. A "1" in the time-series denotes the event of sending an email at that particular time-coordinate. One can infer a candidate path for information flow in this representation by composing a path out of temporally ordered events. One such path $(E \rightarrow C \rightarrow D)$ is marked separately in the figure. If needed, one can also incorporate a notion of "waiting" in these paths to prevent undervaluation of candidate paths. More of this is covered in Chap. 4.

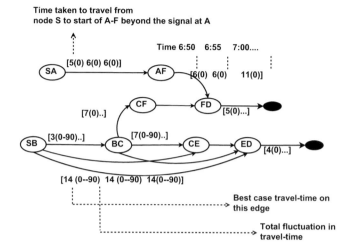

Fig. 3.9 A temporal binary-edged digraph representing road network data. Some travel-time information is not shown in figure to retain its clarity

3.2.2 Temporal Digraphs for Holistic Properties and Turn Delays

Figure 3.9 illustrates a temporal digraph adapted to represent turn delays and holistic properties. The figure models some of these properties for the road network shown in Fig. 3.7. For the purposes of modeling turn-delays (non-deterministic properties), we need to consider a dual graph view. Here, the road segments form the nodes and edges represent an allowed turn (or going straight). In this representation, there are multiple ways to fill-in the road network datasets (TD roadmaps and traffic signal information). However, we would focus only on the representation which would make it easy for the typical query algorithms. For instance, several routing query algorithms typically use an adapted form of the Dijkstra's algorithm at some stage. One may recall that Dijkstra's algorithm assigns labels to nodes and works on optimizing the sum of edge costs in a path. Thus, it would be much more cleaner to associate all the travel-time related data with edges only.

The representation shown in Fig. 3.9 does this in the following way. The data assigned to an edge consists of the following two values: (1) time required to reach the starting point of the next road segment (after the signal) from the starting node of the previous segment without any delay at signal; and (b) worst case delay at the signal. For instance, in the figure, the cost of the edge SA-AF is 5 (0) time units for a 6:50am departure. This means that it takes 5 time units to do the following: travel on road segment SA (after the signal at S, if any), don't wait at intersection A, and then reach the starting point of segment AF. And there is usually no waiting at intersection A.

Table 3.1 Details of edges representing holistic properties

Edge in the digraph	Journey represented
SB-BC	Start at S then: travel S-B (3 min at 6:50am) to the beginning of segment B-C (after SG1) with delay of 0–90 s at 6:53am
SB-CE	Start at S then: travel S-B (3 min at 6:50am) + delay of 0–90 s at SG1 at 6:53am + travel B-C (7 min) to beginning of edge C-E without any wait at SG2
SB-ED	Start at S then: travel S-B (3 min at 6:50am) + delay of 0–90 s at SG1 at 6:53am + travel B-C (7 min) + no delay at SG2 + travel C-E (4 min) to reach the beginning of edge E-D without any wait at SG3
BC-ED	Start at B then: travel B-C (7 min at 6:50am) + delay of 0–90 s SG2 at 6:57am + travel C-E to reach the beginning of edge E-D without any wait at SG3

In figure, signal delays are mentioned in "( )" beside the travel-time information written on the edges. Depending on the application problem, the routing algorithm may either choose to consider the best possible (no-waiting), worst-case possible (90 s) or some statistic (e.g., average of 0 and 90) on the delay.

Representing Holistic Properties A holistic property measured on a sequence of road segments is represented using a series of directed edges. Consider the case of coordinated traffic signals along the route S-B-C-E-D (conceptually shown in the Fig. 3.5) with signals at intersections B (SG1), C (SG2), and E (SG3). This is represented using the following four edges in the digraph: (a) SB-BC, (b) SB-CE, (c) SB-ED and (d) BC-ED. Table 3.1 illustrates the journeys corresponding to these edges. Computing the total travel-time on these edges is slightly non-trivial. Basically, for each of these edges (representing the holistic property), we need to compute the following two values: (1) best case travel-time and (2) worst case travel-time. Using these two we would compute the total fluctuation possible. And, of course, one must consider the travel-time only in Lagrangian reference frame. Following is an example which highlights a corner case to consider while computing the total fluctuation. Consider a case where the edge (journey) under consideration consists of two road segments XY and YZ with signals s1 and s2 at road intersections Y and Z respectively. Signals s1 (max delay 2 min) and s2 (max delay 2 min) are coordinated. For this edge, we need to compute the best case (and the worst case) time taken to travel on a journey which starts at node X, passes through segment XY, waits at s1, then passes through YZ and finally, no wait at s2 (due to signal coordination) to reach the next segment.

In this example, suppose that travel-time on XY at 7:00am is 5 min; the traveler would reach s1 at 7:05am. Now, if the travel-time on edge YZ changes from 7 to 9 min at 7:06am (mathematically possible); then the worst case travel time would be 16 min (5+2+9), whereas the best case travel-time would be 12 min. Consequently, for this edge, we would store best case travel-time (12 min) and a fluctuation of 4 min (2 min due to signal delay and an additional 2 min due to change in typical travel time).

Zero Outdegree Points Special provisions need to be made for "road intersections" which have zero out degree (i.e., no roads going out). Otherwise, we would not be able to represent the travel-time on the segments leading to the particular intersection. Intersection D in Fig. 3.7 was one such example. In order to represent the travel-time on road segment FD and ED, we added dummy nodes (shown in dark shade) in Fig. 3.9. Now the edge connecting FD with the dummy node would have the travel-time information of segment FD. Similar is the case with segment ED.

3.3 Conclusion

Spatio-temporal graphs can be viewed from either an Eulerian or a Lagrangian perspective. Lagrangian perspective is more suitable for studying information flow in social interaction data and navigational queries (on road network datasets). Traditional conceptual models for spatio-temporal graphs include the popular time-expanded graphs which can model decomposable properties. The more recently proposed Lagrangian Xgraphs can represent both decomposable and holistic properties. Both time-expanded graphs and Lagrangian Xgraphs put Lagrangian reference frame upfront. One can define their "Eulerian version" in an analogous way. From an efficiency perspective, both time-expanded graphs and Lagrangian Xgraphs can be compressed into space efficient data structures which can be easily used by the query algorithms.

Chapter 4
Fastest Path for a Single Departure-Time

Finding fastest paths in spatio-temporal graphs have several applications. From the perspective of urban road networks, this problem has applications in navigational systems. For example, one may use algorithms developed for this problem to answer questions like, "what is the fastest path to airport from the University for a 9:00am departure?" With a slight modification of these algorithm, one could also answer the following question: "when is the latest I should depart from my home in order to reach the airport by 5:00pm?" In case of social networks, these algorithms could be used in computing temporal generalizations of shortest path based centrality metrics (e.g., closeness and betweenness centrality).

Outline of the Chapter The rest of the chapter is organized as follows. Section 4.1 provides a formal problem definition and discusses few important considerations. In Sect. 4.2, we provide an adaptation of Dijkstra's algorithm for computing fastest paths in temporal digraph representations of the datasets. A* based search for the Fastest path problem is covered in Sect. 4.3. Section 4.4 covers concepts relating to bi-directional search. In Sect. 4.5, we present adaptations of the Fastest Path algorithm for computing temporal generalizations of centrality metrics.

4.1 Problem Definition

We define the fastest path problem by specifying input, output, objective function and constraints.

Inputs

1. A spatio-temporal graph represented as a temporal digraph $G = (V, E)$, where V is the set of vertices, and E is the set of directed edges. Each edge $e \in E$ is associated with a cost function δ, which gives the cost of the edge as function of time.

© Springer International Publishing AG 2017
V.M.V. Gunturi, S. Shekhar, *Spatio-Temporal Graph Data Analytics*,
https://doi.org/10.1007/978-3-319-67771-2_4

2. A source s and a destination d pair where $\{s, d\} \in V$.
3. A time instant τ.

Output The output is a route, P_{sd}, between s to d.

Objective Function Total cost of P_{sd} is least among all the paths (between s and d) for time τ.

Scope In this chapter we assume that the cost function δ is represented as a time series. Nevertheless, the techniques described in this chapter can be trivially modified to work with continuous functions as well.

4.1.1   Important Considerations

Here, we provide few important aspects to consider while using the output of the fastest path problem. Its important to note that the algorithms presented in the following sections retain their correctness (at the computational level) irrespective of these considerations. These are just important to retain the semantic meaning of the results.

Time-Dependent Cost δ and Reference Frame Depending on the dataset under consideration, the time dependent cost function δ of the edges would obviously mean different things and thus, would call for different reference frames. In case the temporal digraph is representing urban road network (Sect. 3.2 in Chap. 3) datasets (e.g., temporally detailed roadmaps and traffic signal information), then the cost function would be carrying the concept of travel-time. Consequently, for any navigational application of the problem, one must follow only the Lagrangian reference frame. Similar would be the case for email datasets, or for that matter any other social network dataset which captures information flow among individuals. However, if the cost function δ represents something like a time dependent strength of social ties, then one would have to follow the Eulerian reference frame.

Interpreting the Cost Function δ in Road Network Datasets While working with travel-time data in road network datasets, we need to consider the concepts of FIFO and non-FIFO behavior in road networks. FIFO behavior dictates that, in any journey (consisting of any number of edges), a person starting the journey earlier would reach the tail node of the journey earlier, i.e., no overtaking is allowed. However, such is not the case with non-FIFO travel-time. For instance, in Fig. 4.1a, travel-time on edge (A,B) is considered to be non-FIFO as one can reach node B early if he/she starts from node A at time $t = 2$ instead of $t = 1$. Non-FIFO behavior can occur in case of public transport services with different kinds of bus services (e.g., regular vs express). Note that even if one edge exhibits non-FIFO behavior, the digraph is said to be non-FIFO in nature.

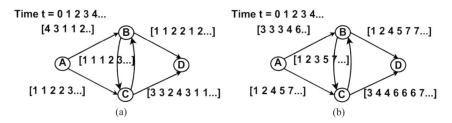

Fig. 4.1 Sample temporal digraphs. (**a**) Digraph with travel-time. (**b**) Digraph with earliest arrival time

Non-FIFO nature of the digraph raises the concept of *waiting* at intermediate nodes to get a quicker paths. Popular literature [57] has conjectured that finding fastest paths in a non-FIFO dataset where the routes are not allowed to have any waiting at the intermediate nodes may be a NP Hard problem. This is due to the fact that by forbidding waiting, we render the fastest paths to go beyond the traditional expectation of being simple (i.e., no nodes are repeated). This precludes the use of Dijkstra's style of enumeration as Dijkstra's does not revisit a node after it is closed.

Non-FIFO behavior can still be incorporated into a Dijkstra's style of enumeration if unrestricted waiting is allowed at the intermediate nodes of the path. This is achieved by converting the travel-time information associated with an edge into *earliest arrival times* [24, 57]. This is a two step process. First, the travel time information is converted into arrival time information. Second, the arrival time information is converted into earliest arrival time information. The second step captures the possibility of arriving at an end node earlier by waiting (non-FIFO behavior). For example, in the temporal digraph shown in Fig. 4.1a, the travel-time series of edge (A, B) is [4 3 1 1 2 ...]. In the first step, we convert it into arrival time series. This is done by adding the time instant index to the cost. For example, if we leave node A at times $t = 0, 1, 2, 3 \ldots$, we would arrive at node B at times $t = 4 + 0, 3 + 1, 1 + 2, 1 + 3, 2 + 4 \ldots$. Therefore, the arrival time series of (A, B) would be [3 4 3 4 6 ...]. The second step involves comparing each value of the arrival time series to the value to its right in the arrival time series. A lower value to its right means that we can arrive at the end node earlier by waiting. Consider the arrival time series of (A, B) = [4 4 3 4 6 ...]. Here, the arrival time for $t = 0$ and $t = 1$ is 4 (which is less than the arrival time for $t = 2$). Therefore, the earliest arrival time on edge (A, B) for times $t = 0$ and $t = 1$ would be set to 3 (i.e., we wait for 2 and 1 time units at node A). This process is repeated for each value in the time series. At the end of the second step, the earliest arrival time series of edge (A, B) would be [3 3 3 4 6 ...]. The algorithm would store the information relating to waiting at times $t = 0$ and $t = 1$ in its data structures. Note that in absence of non-FIFO behavior, the there no change in Step 2. In other words, the earliest arrival time value for a time t would always be equal to whatever was its value after Step 1. Figure 4.1b shows the digraph from Fig. 4.1a after the earliest arrival time series transformation.

Interpreting the Cost Function δ in Social Network Datasets Unlike the road network datasets, social network datasets don't carry a notion of non-FIFO explicitly. However, the concept of *waiting* still comes up in an indirect sense. For example, consider an instance of email dataset for the purposes of studying information flow. In such a dataset, consider a case where an individual A sends an email to B at time $t = 1$ and node B sends another email to individual C at time $t = 5$. Now, $A \rightarrow B \rightarrow C$ may be still be considered a potential path for information flow despite a wait of 4 time units at node B. Depending on the application at hand, one may choose an upper bound on allowed waiting time (or some kind of a decay function). The intuition behind this being that likelihood of a node to forward a piece of information (or meme) decreases with time [35, 65, 66], and probably reduces to zero beyond a certain time. Given such a parameter as input, it is possible to develop a scheme—analogous to the previously described earliest arrival time transformation—to create a new cost function (from δ) for each edge (u, v) which, given a time t, returns the amount of time that the information must wait on the node u before it can be transferred to node v along the edge (u, v).

4.2 Dijkstra's Algorithm for the Fastest Path Problem

Algorithm 1 presents a pseudocode of Dijkstra's algorithm which has been adapted to work with temporal digraphs. The algorithm determines a fastest path between the designated source and the destination node for the time $t = \tau$. Notice that the Algorithm uses Lagrangian reference frame (Line 8, Line 11 and Line 12). But, it can be trivially modified to consider the Eulerian reference frame.

Algorithm 1 Adapted Dijkstra's for computing fastest paths at time τ

  1: Initialize a priority queue
  2: Insert source node, label(source)=τ /* label(v) is the time at which current best path arrives at
     v.*/
  3: $u \leftarrow$ ExtractMin
  4: **while** u is not the destination node **do**
  5:     Add u to the list of closed nodes
  6:     **for all** neighbors v of u **do**
  7:         **if** v is not in the priority queue **then**
  8:             label(v) $\leftarrow \delta_{(u,v)}(label(u))$ /* Assuming δ contains earliest arrival times.*/
  9:             predecessor(v) $\leftarrow u$
 10:             Insert v into the priority queue
 11:         **else if** label(v) $> \delta_{(u,v)}(label(u))$ **then**
 12:             label(v) $\leftarrow \delta_{(u,v)}(label(u))$
 13:             predecessor(v) $\leftarrow u$
 14:         **end if**
 15:     **end for**
 16:     $u \leftarrow$ ExtractMin
 17: **end while**
 18: Output the source-destination path.

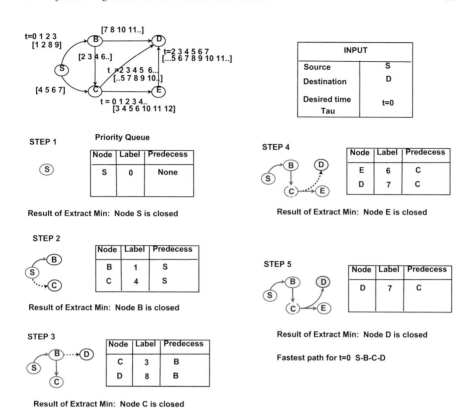

Fig. 4.2 Execution trace of adapted Dijkstra's on temporal digraphs

Execution Trace Figure 4.2 provides an execution trace of the Algorithm 1 for a sample temporal digraph. The input temporal digraph is shown on top left of figure. Here, node S is the source, D is the destination and, we are interested in computing a fastest path for the time $t = 0$ (i.e., $\tau = 0$).

It is important to note that the pseudocode given in Algorithm 1 can be trivially modified to compute the breadth first search tree on a temporal digraph. For that, δ would simply consist of 1's or $+\infty$'s and the algorithm would run until it discovers all nodes. Depending on the choice of the reference frame (Lagrangian vs Eulerian), the edges in the tree can either all have the same time-coordinate or span across different times (Lagrangian reference frame). A breadth first search tree with Lagrangian reference frame can be used for studying the disease spread problem (over a social network) mentioned in Chap. 1. For example, if the disease is such that people may become its carries without showing symptoms or have only mild symptoms which can go undetected, then one would have to perform a breadth first search (BFS) (with Lagrangian reference frame) over the social interaction data, starting from the infected person at the time when he/she started showing symptoms. Nodes in this tree would be considered as potential carries. Depending on the nature of disease, this BFS tree can be limited upto a certain time instant.

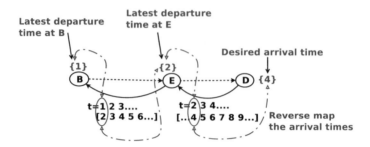

Fig. 4.3 Reverse search on temporal digraphs

4.2.1 Latest Departure Path Problem

Dijkstra's algorithm can also be adapted to answer the following question on temporal digraphs: "when should I depart to reach my destination by 5:00pm?" In this problem, the input consists of a temporal digraph, a source, a destination and a desired arrival time ω at the destination. Here, the cost function δ of edges is assumed to be in form of earliest arrival times.

As expected, in this problem, we would be working with a reverse of the original temporal digraph where an edge in the original digraph (u, v) is replaced by its reverse (v, u). However, the cost functions (δ) of edges in the original digraph need to be used with certain thought as explained next. As a example, consider the simple temporal digraph shown in Fig. 4.3 containing just three nodes B, E and D. The figure shows the original edges with dotted lines and reverse edges with bold lines. The figure also shows cost functions δ of the original edges (B, E) and (E, D). In the problem instance illustrated in Fig. 4.3, we are interested in determining the latest departure time at node B for arriving at node D by time $t = 4$. Obviously, the optimal path would be $B - E - D$, given that there are no other paths between B and D. The latest departure-time, in this example, would computed as follows. Starting from node D at time $t = 4$, we ask the following question: "When is the latest I can depart from E in order to arrive at D by $t = 4$?" This obtained by reading the δ of the edge (E, D) in an inverse manner. For time $t = 2$, $\delta_{E,D}$ has a value of 4 (the desired arrival time at D). We continue this process along each edge of the path $D - E - B$ to get the latest departure-time of $t = 1$ at the node B.

Unlike the example shown in Fig. 4.3, it may not be always possible to inverse map the desired arrival times. For example, consider the problem instance shown in Fig. 4.4. Here, the desired arrival time of 7 at D cannot be mapped back (along the edge (E, D)) to a departure-time at node E. In such cases, we have following two options. First, we may map it to the latest departure-time where $\delta_{E,D}$ value is less than 7, which in this case happens to be time $t = 4$ (as shown in Fig. 4.4). Second, we may map it to $-\infty$, implying that the path does not exist. The choice would

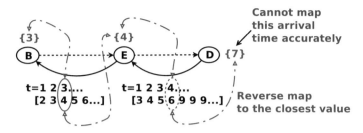

Fig. 4.4 Reverse search on temporal digraphs: closest mapping

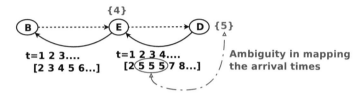

Fig. 4.5 Reverse search on temporal digraphs: ambiguous case

depend on the scope of the problem at hand. Note that in some cases mapping to $-\infty$ would almost be mandatory. Consider a case where we are trying to find a inverse mapping of the arrival time 2 along the edge (E, D). This is physically impossible. In such cases, we would have to map it to $-\infty$ to maintain mathematical sanity of the algorithm. Implicitly it means that the length of the path for this time is ∞, which is as good as saying that the path does not exist at this time.

Figure 4.5 illustrates another case which requires special attention. Here, the arrival time of $t = 5$ at node D faces the case of ambiguous mapping along the edge (E, D). Arrival time of $t = 5$ can be mapped back to departure-times 4, 3 or 2 at node E. But following our principle of using only the latest departure-times, we would map it only 4 at node E. There is a certain amount of waiting in the departure-times 2 and 3.

With the intention of keeping the discussion focused, we are not providing a pseudocode for computing the latest departure path. This algorithm would in essence be very similar to Algorithm 1 with just following three changes:

1. Use reverse of input temporal digraph.
2. Relax the edges using the schemes of inverse mapping over δ as described previously.
3. The priority queue needs to store the absolute value of path lengths (in terms of total travel-time) rather than the node labels. Note that now the node labels would contain the latest departure-time to reach the destination at the desired arrival time ω (along the fastest path).

4.3   A* Search for the Fastest Path Problem

A* based searches are an essential tool to speed up path finding algorithms on large networks. Consider the sample network shown in Fig. 4.6. Here each edge is bidirectional and has unit cost. In this network, an instance of Dijkstra's based algorithm starting from node S may first close all the nodes in the graph before determining a path to the destination node D. This is because of the greedy nature of the algorithm. If the optimal path length is x units, then Dijkstra's would first determine all the shortest paths which are of length less than x (and $= x$ as ties are broken arbitrarily in the algorithm). The same issue comes up again when Dijkstra's algorithm is adapted to temporal digraphs (Algorithm 1).

A* based searches are based on the concept of what are known as the *distance-estimator functions*. Given any node v, the *distance-estimator function* estimates the shortest path length to the destination from the node v. These estimates are used along with the regular labels (denoting the length of the best path found so far) in the Dijsktra's to prune the search space. However, in order to retain the correctness of the result, a viable *distance-estimator function* should have the following properties: (a) admissibility and (b) monotonicity. Admissibility states that for any node v, the distance-estimator should never overestimate the distance to the destination. Monotonicity states that, for given any edge (x, y), distance-estimate$(x) \leq Length(x, y) +$ distance-estimate(y).

Traditionally, in spatial graphs, Euclidean distance between v and the destination was used as the distance-estimator function. One can easily verify that Euclidean distance satisfies both the admissibility and the monotonicity requirement. Researchers in the area of spatio-temporal computing [10, 14, 21] have explored other distance-estimator functions which could take advantage of some special features of spatio-temporal graphs. We now describe two of them.

4.3.1   Using Lower Bound on Travel-Time

In this technique, given a temporal digraph, we first derive a *lower bound graph* (*LG*) by replacing the cost function (δ) of each edge with its single minimum value (*minδ*)

Fig. 4.6 Sample road network illustrating the need for A* based search

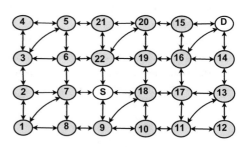

over the entire time horizon of interest. Now, the value of distance-estimator ($\mathcal{L}(v)$) at a node v would be the length of the shortest path between v and the destination in the aforementioned lower bound graph. Admissibility of this distance-estimator can be trivially established using just the construction of lower bound graph. Lemma 4.1 establishes the monotonicity of this distance-estimator function.

Lemma 4.1 *Given an* lower bound graph *which is constructed from a temporal digraph by replacing δ of each edge with its minimum value (minδ) over the entire time horizon, we have following:*

- *A distance-estimator $\mathcal{L}(v)$ at a node v, which returns the shortest distance between v and destination d in the lower bound graph, is monotonic.*

Proof Without loss of generality, assume a time t for which this property has to be established. We need to show that $\mathcal{L}(x) \leq \delta_{(x,y)}(t) + \mathcal{L}(y)$, $\forall (x, y)$ *and* t. We have two cases. Case (a): $\delta_{(x,y)}(t) = min\delta_{(x,y)}$. And case (b): $\delta_{(x,y)}(t) > min\delta_{(x,y)}$.

Case (a) In this case, we have two options, either the shortest path (in the lower bound graph) between x and d goes through node y, or it does not. In case the shortest path from x goes through y, then $\delta_{(x,y)}(t)$ can either be greater than or equal to the shortest distance between x and y. Either way, $\mathcal{L}(x) \leq \delta_{(x,y)}(t) + \mathcal{L}(y)$. We will get equality when the shortest path goes through the edge (x, y) or is of length equal to the length of the edge (x, y).

Now consider the case when the shortest path (in the lower bound graph) from x *does not* go through y. Without loss of generality, assume that the shortest path as determined by $\mathcal{L}(x)$ and $\mathcal{L}(y)$ have same nodes $d, v_1, v_2, \ldots, v_k$ ($k \geq 0$). Now, consider the lengths of sub-paths which are different. Clearly, length of the shortest path from x to the last uncommon node p would have to be less than or equal to the term: $\delta_{(x,y)}(t)$ + length of shortest path from y to the last uncommon node q. Otherwise, the shortest path from x would have actually gone through (x, y) which would be a contradiction to original assumption. Thus, proving the monotonicity in this case as well.

Case (b) Proof for this case can be easily derived from case (a) by noting the fact that substituting $min\delta_{(x,y)}$ in the right hand side of $\leq$ with a larger value $\delta_{(x,y)}(t)$ would not change the inequality.

4.3.2 Using a Hierarchical Lower Bound on Travel-Time

While the concept of using lower bound on travel-time as a distance-estimator ensures the correctness of the solution, its implementation, however, requires us to precompute and store the lower bound travel-time between all pairs of nodes in the underlying graph. To this end, researchers [14] have proposed a technique which is more efficient.

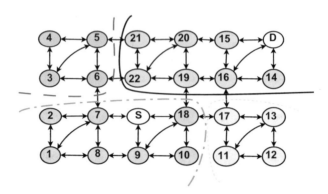

Fig. 4.7 Sample road network partitions for hierarchical lower bound on travel-time (best in color)

The key idea here is to spatially partition the nodes of the lower bound graph into a disjoint set of partitions. There are several ways to create this partitioning, some of the well accepted techniques include [14, 32, 59]. In addition to these, one can also come up intuitive partitioning techniques which use the well defined hierarchy of roads consisting of interstate freeways and arterial roads. We could first divide a large region using the top level roads such as the interstate freeways. Then, the sub-regions could be divided using the next level of roads such as the arterial roads. This could recursively proceed until each resulting partition attains a "manageable size".

As an example consider the sample partitioning shown in the Fig. 4.7. Here, the network was partitioned into four parts. The figure shows each partition is shown in a different color. There are three key concepts in this technique.

1. **Border Nodes of a partition** p_i**:** A border node $b_i^{p_i}$ of a partition p_i is defined as a node which has at least one edge to a node of a different partition p_j. This can either be a outgoing or an incoming edge. For instance, in Fig. 4.7, nodes 5 and 6 are the border nodes for the red partition. Similarly, nodes 7 and 18 are the border nodes for the green partition.
2. **Inter-partition lower bound travel-time** (Γ)**:** Given a partition p_i with its set of boundary nodes $\{b_1^{p_i}, b_2^{p_i}, \ldots, b_k^{p_i}\}$, and another partition p_j with its set of boundary nodes $\{b_1^{p_j}, b_2^{p_j}, \ldots, b_m^{p_j}\}$. The inter-partition lower bound travel-time between p_i and p_j ($\Gamma(p_i, p_j)$) would be the minimum of the pair-wise shortest path lengths (in the lower bound graph LG) between all pairs $(b_x^{p_i}, b_y^{p_j})$, $\forall x \in [1k]$ *and* $\forall y \in [1m]$. In other words:
   $\Gamma(p_i, p_j) = Min\{SPLen(b_x^{p_i}, b_y^{p_j}), \forall x \in [1k]$ *and* $\forall y \in [1m]\}$, where $SPLen(b_x^{p_i}, b_y^{p_j})$ denotes the shortest path length between $b_x^{p_i}$ and $b_y^{p_j}$ in the lower bound graph LG.
3. **Intra-partition lower bound travel-time** (Ω)**:** This is defined between a node v and its corresponding partition p_i. $\Omega(v, p_i)$ is set to the minimum of the pair-wise shortest path lengths in the lower bound graph between all pairs $(v, b_x^{p_i})$, $\forall b_x^{p_i}$ in the set of boundary nodes of p_i. In other words:

$\Omega(v, p_i) = Min\{SPLen(v, b_x^{p_i}), \forall b_x^{p_i}$, where $SPLen(v, b_x^{p_i})$ denotes the shortest path length between v and $b_x^{p_i}$ in the LG. $\Omega(p_i, v)$ is defined in a similar way. Note that in road networks $\Omega(v, p_i) \neq \Omega(p_i, v)$

Definition 4.1 Distance-estimator function $\mathcal{H}$ Given a node v and destination node d, the value of the distance-estimator $\mathcal{H}(v)$ is computed as follows:

Case (a) If v and d are in the same partition then, $\mathcal{H}(v)$ is the shortest path length between v and d in the lower bound graph.

Case (b) If v is in partition p_i and d is in partition p_j, then $\mathcal{H}(v) = \Omega(v, p_i) + \Gamma(p_i, p_j) + \Omega(p_j, d)$.

The distance-estimator function $\mathcal{H}$ used in this scheme is formally given in Definition 4.1. The admissibility and the monotonicity of this distance-estimator can be established using arguments similar to the ones used in Lemma 4.1. The key fact to note is that, given the nature of construction of this distance-estimator, one can easily establish that $\mathcal{H}(v) \leq \mathcal{L}(v), \forall v$ and all destinations. Algorithm 2 presents a pseudocode of A* based search for the fastest path problem. In this algorithm, label of node v which goes into the priority queues consists of sum of the following two terms: (a) current upper bound on travel-time through the best path found so far (denoted as $SD()$ in the pseudocode), and (b) $\mathcal{H}(v)$ or $\mathcal{L}(v)$ depending on the choice of the distance-estimator used in the application.

Algorithm 2 A* search for computing fastest path at time τ

1: Initialize a priority queue
2: Initialize $SD(source) = \tau$ Assuming δ contains earliest arrival times.
3: Insert source node into queue, label(source)=$SD(source) + \mathcal{H}(source)$
4: $u \leftarrow$ ExtractMin
5: **while** u is not the destination node **do**
6:     Add u to the list of closed nodes
7:     **for all** neighbors v of u **do**
8:         **if** v is not in the priority queue **then**
9:             $SD(v) \leftarrow \delta_{(u,v)}(SD(u))$
10:             label$(v) \leftarrow SD(v) + \mathcal{H}(v)$
11:             predecessor$(v) \leftarrow u$
12:             Insert v into the priority queue
13:         **else if** $SD(v) > \delta_{(u,v)}(SD(u))$ **then**
14:             $SD(v) \leftarrow \delta_{(u,v)}(SD(u))$
15:             label$(v) \leftarrow SD(v) + \mathcal{H}(v)$
16:             predecessor$(v) \leftarrow u$
17:         **end if**
18:     **end for**
19:     $u \leftarrow$ ExtractMin
20: **end while**
21: Output the source-destination path.

4.4   Bi-directional Search for the Fastest Path Problem

Apart from A* based search, one can also use bi-directional search over Dijkstra's for pruning the search space of candidate routes without compromising on the correctness of the solution. However, a bi-directional search over temporal digraphs presents following two challenges:

1. While the forward search may start from source at the desired departure-time given in the input, we don't have similar information regarding the "arrival-time" at the destination. Ideally, the backward search should start from the destination at this "arrival-time."
2. Devising a suitable termination condition for bi-directional search in temporal digraphs is non-trivial.

For addressing the first challenge, research literature [14, 52, 53] has suggested to execute the backward search on a reverse of the lower bound graph (refer Sect. 4.3 for definition). In this reverse graph every edge (u, v) in the original lower bound graph is replaced by its reverse (v, u). In these algorithms [14, 52, 53], one could imagine the backward search as something of a guide to the forward search by pruning the space to be explored by the forward search.

Addressing the second challenge of designing a suitable termination condition is slightly non-trivial and is explained next. Terminating when both the forward and the backward searches close a common node may lead to sub-optimal results. Figure 4.8 shows an example illustrating this concept. This kind of termination condition can certainly not guarantee optimal results in static graphs. In temporal digraphs, this termination may guarantee optimality, but only in very few cases (more details in Chap. 6). We now describe another termination condition (proposed in [14]) for bi-directional search in temporal digraphs which guarantees correctness for the stated forward and backward searches.

Consider a forward search from the source and a backward search from the destination. As explained earlier, the forward search would be explore the input temporal digraph, while the backward search would be working on the reverse of the lower bound graph (created out of the input temporal digraph). Internally, both forward and the backward searches would be similar to an instance of the Dijsktra's with nodes being assigned labels denoting the length of the current best path found so far from the source (in case of forward search) and destination (in case of backward search) respectively. To be more precise, the forward search would be similar to an instance of adapted Dijkstra's for a departure-time τ on the input temporal digraph (refer Algorithm 1). Whereas, the backward search would be a simple Dijsktra's on the reverse of the lower bound graph.

In the bidirectional algorithm, both the forward and the backward search would be executed in an interleaved manner. Basically, the algorithm would process one "ExtractMin" from each of the searches in one iteration. This process continues until both forward and backward searches close a common node. As an example, let the first common node to be closed by both the searches be v. Now consider the path source-v-destination obtained by stitching the paths determined by the forward and

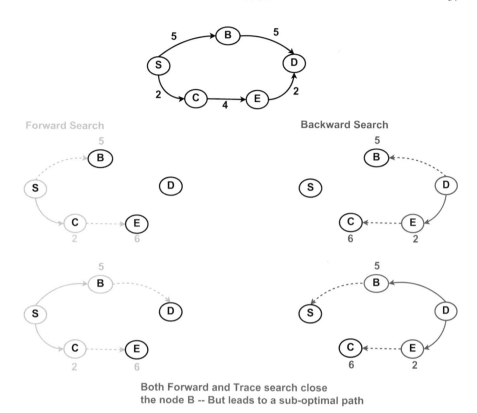

Fig. 4.8 Sample illustrating an incorrect termination condition in bidirectional search

the backward search. Length of this path in the temporal digraph would certainly be an upper bound on the fastest path between source and destination for the time τ. A more tighter bound would be the sum of the lengths of the following: (a) path determined by the forward search and, (b) fastest path between v and destination in the temporal digraph for the time corresponding to when the forward search arrived at node v. Pseudocode given in this monograph uses the latter upper bound.

Once an upper bound on the fastest path length between source-destination is known, the algorithm becomes more careful in exploring the search space. This is done by continuing the forward and backward search with a slightly different termination condition; in the first part, the termination condition was when both searches closed a common node. In second part, both the searches are continued until the all node labels in the queue of the backward search are greater than the current best upper bound (on source-destination path) found so far. During this process, both the searches are likely to close more common nodes. With each of these common nodes, the algorithm tests if the upper bound can be improved. In the final part of the algorithm, if the destination has not yet been closed, only the forward search continues. However, now the forward search would only consider the nodes visited so far by the backward search.

A pseudocode of the bi-directional search algorithm is given in Algorithm 3. Lines 7–28 correspond to the first part of the algorithm, where both the forward and the backward searches are executed in an interleaved manner until both of them close a common node. Following this, the algorithm computes an upper bound in line 30 based on common node x. Lines 31–35 correspond to the second part, where the searches are continued until all nodes in the priority queue of the backward search are greater than the current upper bound (updated in lines 33–34).

Algorithm 3 Bi-directional Dijkstra's for computing fastest paths at time τ

1: Initialize a priority queue F_{pq} for forward search /*Runs on input temporal digraph.*/
2: Initialize a priority queue B_{pq} for backward search /*Runs on reverse of lower bound graph*/
3: Insert source node into F_{pq}, label(source)=τ /*label(v) is the time at which current best path arrives at v.*/
4: Insert destination node into B_{pq}, label(destination)=0
5: $f \leftarrow$ ExtractMin from F_{pq} and $b \leftarrow$ ExtractMin from B_{pq}
6: Add f to forward closed nodes set ($Fcol$) and b to backward closed nodes set ($Bcol$)
7: **while** $Fcol \cap Bcol = \emptyset$ **do**
8:     **for all** neighbors v of f **do**
9:         **if** v is not in F_{pq} **then**
10:            label(v) $\leftarrow \delta_{(f,v)}(label(f))$ /*Assuming δ contains earliest arrival times. */
11:            predecessor(v) $\leftarrow f$
12:            Insert v into the priority queue
13:        **else if** label(v) $> \delta_{(f,v)}(label(f))$ **then**
14:            label(v) $\leftarrow \delta_{(f,v)}(label(f))$
15:            predecessor(v) $\leftarrow f$
16:        **end if**
17:    **end for**
18:    **for all** neighbors u of b **do**
19:        **if** u is not in B_{pq} **then**
20:            label(u) $\leftarrow label(b) + edge_cost(b,u)$ and predecessor(u) $\leftarrow b$
21:            Insert u into the priority queue
22:        **else if** label(u) $> (label(b) + edge_cost(b,u))$ **then**
23:            label(u) $\leftarrow label(b) + edge_cost(b,u)$ and predecessor(u) $\leftarrow b$
24:        **end if**
25:    **end for**
26:    $f \leftarrow$ ExtractMin from F_{pq} and $b \leftarrow$ ExtractMin from B_{pq}
27:    Add f to forward closed nodes set ($Fcol$) and b to backward closed nodes set ($Bcol$)
28: **end while**
29: $x \leftarrow$ Common node closed by both searches
30: src-dest ubound $\leftarrow$ Forwardsearch label(x) + FastestPathLen(x,dest,forward label(x))
31: **while** (top of B_{pq} queue $<$ src-dest ubound) ‖ destination not closed by forward **do**
32:    Continue the forward and the backward search similar to lines
33:    After each Step of forward and backward search check $Fcol \cap Bcol$
34:    For any new common node y closed by both searches check if the src-dest ubound can be improved
35: **end while**
36: **while** destination not closed by forward search **do**
37:    Continue the forward but consider only the nodes visited so far by the backward search.
38: **end while**
39: Output the source-destination path.

4.5 Adapting Fastest Path Algorithm for Centrality Metrics

There have been several works [28, 30, 40, 41, 61] along the lines of extending the traditional shortest path based centrality metrics such as Betweenness [3, 18, 19] and Closeness [19] for temporal digraphs. Variant of Dijkstra's algorithm discussed in Sect. 4.2 can also be easily adapted to compute shortest path based centrality measures such as Betweenness and Closeness on temporal digraphs. Amongst these two adaptation for Closeness is trivial given the nature of the metric which is just a inverse of sum of lengths of fastest paths. In this monograph, we only discuss betweenness centrality given its non-triviality.

Though there have been several temporal generalizations of betweenness centrality, with the intention of maintaining simplicity, we would be using the generalization given in Eq. (4.1). In Eq. (4.1), $C_B(v)[t]$ denotes the betweenness centrality of a node v at time t. In the equation, $\sigma_{sd}[t]$ refers to the number of shortest paths between node s and node d at time t, and $\sigma_{sd}(v)[t]$ refers to the number of shortest paths between node s and node d that pass through node v at time t. Here, the concept of "shortest path between node s and node d at time t" is interpreted according to Definition 4.2. This interpretation follows the Lagrangian reference frame used to model flow of information.

$$C_B(v)[t] = \sum_{\forall s \neq v \neq d \in V} \frac{\sigma_{sd}(v)[t]}{\sigma_{sd}[t]} \tag{4.1}$$

Definition 4.2 (Temporal Shortest Path:) A path $P_{sd} = s, x_1, x_2, x_3 \ldots, x_k, d$ between s and d, where s and x_1 interacted at time t, is considered to be a shortest path for time t if it has following two properties.

- Any individual (node of path) may appear only once in the path.
- All the nodes in P_{sd} should be visited in a *temporally ordered* sense. This means that if (x_i, x_j) and (x_j, x_k) are two consecutive edges in P_{sd} corresponding to the interactions at times $t_{x_i x_j}$ and $t_{x_j x_k}$, then $t_{x_i x_j} < t_{x_j x_k}$.
- If there is another path $Q_{sd} = s, y_1, y_2, y_3 \ldots, d$ between s and d, where s and x_1 interacted at time t, then the interaction time of the edge (y_k, d) is not less than that of (x_k, d), i.e, $t_{y_k d} \geq t_{x_k d}$.

Internally we would use a temporal generalization of path counting technique proposed in [8]. More specifically, the notion of *pair-dependencies* in [8] can be generalized to a temporal case. Equation 4.2 shows this generalization for the metric given in Eq. (4.1). Here, $Dependency_{sd}(v)[t]$ denotes the *temporal pair-dependency* on node v at time t.

$$C_B(v)[t] \quad = \sum_{\forall s \neq v \neq d \in V} \frac{\sigma_{sd}(v)[t]}{\sigma_{sd}[t]}$$

$$= \sum_{s \neq v \neq d \in V} Dependency_{sd}(v)[t] \ where \ Dependency_{sd}(v)[t] \ = \ \frac{\sigma_{sd}(v)[t]}{\sigma_{sd}}$$

$$= \sum_{\forall s \in V} Dependency_{s\cdot}(v)[t] \ where \ Dependency_{s\cdot}(v)[t] \ = \ \sum_{\forall d \in V} \delta_{sd}(v)[t]$$

$$(4.2)$$

Here, a single $Dependency_{s\cdot}(v)[t]$ is computed using Eq. (4.3). This is a temporal generalization of Theorem 6 in [8]. This equation is computed using recursive fashion (using stacks), starting with the last node in a shortest path whose $Dependency_{s\cdot}()$ is initialized to zero. In case we need to study the centrality of an individual over a time interval, Eq. (4.1) would have to be computed repeatedly for all the times in the time-interval of interest. Algorithm 4 provides a pseudo-code for computing the temporal betweenness of all nodes in a given temporal digraph according to Eq. (4.1).

$$Dependency_{s\cdot}(v)[t] \quad = \sum_{w:v \in P_s(w)[t]} \frac{\sigma_{sv}(v)[t]}{\sigma_{sw}[t]} \cdot (1 + Dependency_{s\cdot}(w)[t])$$

$$(4.3)$$

where $P_s(w) = \{u \in V : u \ is \ predecessor \ to \ w \ in \ a \ shortest \ path$

$(According \ to \ Definition \ 4.2) \ from \ s \ for \ time \ t\}$

Pseudocode shown in Algorithm 4 can be trivially modified to compute betweenness centrality under some special cases. For instance, consider a scenario where one is interested in determining the key liaisons for information flowing from sales department to finance department. According to Eq. (4.1), this can be computed by considering only those sources (s in Eq. 4.1) which lie in sales department. In addition, we should consider only those destinations (d in Eq. 4.1) which lie the finance department. Both these changes can be easily incorporated in Algorithm 4. Relevant sources can be considered by changing line 2 of the pseudocode accordingly. For considering only the relevant destinations, we ignore all w's on Line 30 which are not from the finance department.

4.6 Conclusions

There is a rich literature for computing the fastest paths for a single departure-time. Researchers have explored a wide variety of techniques for this problem which include bi-directional and A* based searches. These algorithms can easily modified to compute shortest path based centrality metrics such a closeness and betweenness on temporal digraphs.

Algorithm 4 Adapted Dijkstra's for computing betweenness centrality at time τ

1: Initialize centrality C_B of all nodes $w \in V$ to 0
2: **for all** nodes $s \in V$ **do**
3:      Initialize a Stack S and a Priority Queue PQ
4:      Insert source node, label(s)=τ /* label(v) is the time at which current best path arrives at v.*/
5:      Initialize sigma[w] and Pred[w] to NULL for all nodes x
6:      **while** PQ is not empty **do**
7:          $u \leftarrow$ ExtractMin
8:          Add u to the list of closed nodes
9:          Add PotentialPred[u] to predecessor list of u Pred[u]
10:         sigma[u] $\leftarrow$ sigma[u] + sigma[$Pred[u]$]
11:         Push u onto the Stack S
12:         **for all** neighbors v of u **do**
13:             **if** v is closed and label(u) = $\delta_{v,u}(label(v))$ **then**
14:                 sigma[u] $\leftarrow$ sigma[u] + sigma[v]
15:                 Add v to the predecessor list of u Pred[u]
16:             **else**
17:                 **if** v is not in the priority queue **then**
18:                     label(v) $\leftarrow \delta_{(u,v)}(label(u))$ /* Assuming δ contains earliest arrival times.*/
19:                     PotentialPred[v] $\leftarrow u$
20:                     Insert v into the priority queue
21:                 **else if** label(v) > $\delta_{(u,v)}(label(u))$ **then**
22:                     label(v) $\leftarrow \delta_{(u,v)}(label(u))$
23:                     PotentialPred[v] $\leftarrow u$
24:                 **end if**
25:             **end if**
26:         **end for**
27:     **end while**
28:     Initialize Dependency[v] $\leftarrow 0 \ \forall v \in V$
29:     **while** Stack S is not empty **do**
30:         $w \leftarrow$ Pop S
31:         **for all** $v \in Pred[w]$ **do**
32:             Dependency[v] $\leftarrow$ Dependency[v] + $\frac{Sigma[v]}{Sigma[w]} \cdot (1 + Dependency[w])$
33:             **if** $w \neq s$ **then**
34:                 $C_B[w] \leftarrow C_B[w] + Dependency[w]$
35:             **end if**
36:         **end for**
37:     **end while**
38: **end for**

Chapter 5
Advanced Concepts: Critical Time Point Based Approaches

Given a spatio-temporal (ST) graph, a source, a destination, and a start-time interval, the All-start-time Lagrangian shortest paths problem (ALSP) determines a path set which includes the shortest path for every start time in the interval. The ALSP determines both the shortest paths and the corresponding set of time instants when the paths are optimal.

For example, consider the problem of determining the shortest path between the University of Minnesota and the MSP international airport over an interval of 7:00 a.m. through 12:00 noon. Figure 5.1a shows two different routes between the University and the Airport. The 35W route is preferred outside rush-hours, whereas the route via Hiawatha Avenue is preferred during rush-hours (i.e., 7:00 a.m.– 9:00 a.m.) (see Fig. 5.1b). Thus, the output of the ALSP problem may be a set of two routes (one over 35W and one over Hiawatha Avenue) and their corresponding time intervals.

Challenges ALSP is a challenging problem as the ranking of alternate paths between any particular source and destination pair in the network is not stationary. In other words, the optimal path between a source and destination for one start time may not be optimal for other start times. A naive approach for solving ALSP problem would involve determining the shortest path for each start time in the interval. This leads to redundant re computation of the shortest path across consecutive start times sharing a common solution. Some efficiencies can be gained using a time series generalization of a label-correcting algorithm [42]. This approach was previously used to find best start time [23], and was generalized for ALSP in [27]. However, this approach still entailed large number of redundant recomputations. Few other studies [16, 34] considered the problem of reducing the number of redundant recomputations. However, they were either not scalable or did not have enough reduction (for more details refer [29]).

© Springer International Publishing AG 2017
V.M.V. Gunturi, S. Shekhar, *Spatio-Temporal Graph Data Analytics*,
https://doi.org/10.1007/978-3-319-67771-2_5

Time	Preferred Routes
7:30am	Via Hiwatha
8:30am	Via Hiwatha
9:30am	via 35W
10:30am	via 35W

(b)

(a)

Fig. 5.1 Problem illustration. (**a**) Different routes between University and Airport. (**b**) Optimal times

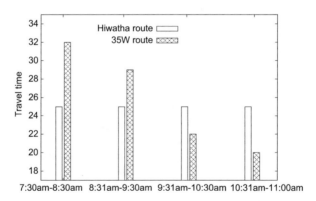

Fig. 5.2 Total travel time of candidate paths

Overview of the Solution Consider again the problem of determining the shortest path between University of Minnesota and MSP international airport over a time interval of 7:30 a.m. through 11:00 a.m. Figure 5.1b shows the preferred paths at some time instants during this time interval, and Fig. 5.2 shows the travel-times for all the candidate paths during this interval.

As can be seen, the Hiawatha route is faster for times in the interval [7:30 a.m. 9:30 a.m.),[1] whereas 35W route is faster for times in the interval [9:30 a.m. 11:00 a.m.]. This shows that the shortest path changed at 9:30 a.m. We define this time instant as *critical time point*. Critical time points can be determined by computing the earliest intersection points between the functions representing the total travel time of paths. For example, the earliest intersection point of Hiawatha route was at 9:30 a.m. (with 35W route function). Therefore, it would be redundant to recompute shortest paths for all times in interval (7:30 a.m. 9:30 a.m.) and (9:30 a.m. 11:00 a.m.] since the optimal path for times within each interval did not change. This approach is particularly useful in case when there are a fewer number of critical time points.

Outline of the Chapter A formal problem definition is presented in Sect. 5.1. A description of the computational structure of the ALSP problem is presented in Sect. 5.2. A critical-time point based algorithm for ALSP is presented in Sect. 5.3. Section 5.4 presents the correctness and completeness proof of the CTAS algorithm. Experimental analysis of the proposed methods is presented in Sect. 5.5. Finally, Sect. 5.6 concludes this chapter.

5.1 Formal Problem Definition

We define the ALSP problem by specifying input, output, objective function and constraints.

Inputs

1. A temporally detailed roadmap dataset represented as temporal digraph $G = (V, E)$, where V is the set of vertices, and E is the set of directed edges.
2. Each edge $e \in E$ is associated with a cost function δ, which is discrete in nature. This function gives the arrival time (or the earliest arrival time in case of non-FIFO; refer Sect. 4.1 in Chap. 4) at the end node of the edge for different departure-times at the starting node of the edge.
3. A source s and a destination d pair where $\{s, d\} \in V$.
4. A discrete time interval λ over which the shortest path between s and d is to be determined.

Output The output is a set of routes, P_{sd}, from s to d where each route $P_i \in P_{sd}$ is associated with a set of start time instants ω_i, where ω_i is a subset of λ.

Objective Function Each path in P_{sd} is the earliest arrival path between s and d for its respective time instants.

We assume that the length of the time horizon over which the ST network is considered is finite. The weight function δ is a discrete time series.

---

[1]Note: an interval (a,b) does not include the end points a and b, $[a,b]$ includes the endpoints a and b, and [a,b) includes a but not b.

Fig. 5.3 Sample input and output of ALSP problem. (**a**) Temporal digraph with arrival times. (**b**) Output of ALSP problem

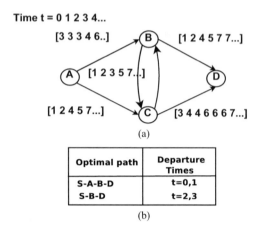

Time t = 0 1 2 3 4...

[3 3 3 4 6..] B [1 2 4 5 7 7...]

A [1 2 3 5 7..] D

[1 2 4 5 7...] C [3 4 4 6 6 6 7...]

(a)

Optimal path	Departure Times
S-A-B-D	t=0,1
S-B-D	t=2,3

(b)

Example Consider the sample temporally detailed roadmap shown in Fig. 5.3a. An instance of the ALSP problem on this network with source A destination D and $\lambda = [0, 1, 2, 3]$, is shown in Fig. 5.3b. Here, path A-C-D is optimal for departure-times $t = 0$ and $t = 1$, and path A-B-D is optimal for start times $t = 2$ and $t = 3$.

5.2   Computational Structure of the ALSP Problem

As mentioned earlier, a key challenge in solving the ALSP problem involves addressing the inherent non-stationarity present in a spatio-temporal graph. Due to this non-stationarity, traditional algorithms developed for static graphs cannot be used to solve the ALSP problem because the optimal sub-structure property no longer holds in case of ST networks. On the other hand, algorithms developed for computing shortest paths for a single start time [16, 23, 39, 43, 57] are not practical because they would require redundant re-computation of shortest paths across the start times sharing a common solution.

We propose a divide and conquer based approach to handle network non-stationarity. In this approach we divide the time interval over which the network exhibits non-stationarity into smaller intervals which are guaranteed to show stationary behavior.[2] These intervals are determined by computing the critical time points, that is, the time points at which the cost functions representing the total cost of the path as a function of time intersect. Now, within these intervals, the earliest arrival path[3] can be easily computed using a single run of a dynamic programming (DP) based approach [23]. Recall our previous example of determining the shortest paths

---

[2]By stationarity, we mean that ranking of the alternate paths between a particular source-destination pair does not change within the interval i.e, there is a unique shortest path.

[3]In this chapter we would use the term shortest path and earliest arrival path interchangeably.

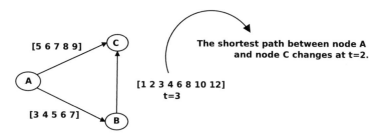

Fig. 5.4 Illustrating non-stationarity

between the university and the airport over an interval of [7:30 a.m. 11:00 a.m.].
Here, 9:30 a.m. was the critical time point. This created two discrete sub-intervals
[7:30 9:30) and [9:30 11:00]. Now, we can compute the ALSP using two runs of a
DP based algorithm [23] (one on [7:30 9:30) and another on [9:30 11:00]).

There are two ways to determine the stationary intervals, either by precomputing
all the critical time points, or by determining the critical time points at run time. The
first approach was explored in [26] for a different problem. Precomputing the critical
time points involves computing intersection points between cost functions of all
the candidate paths. Now, in a real transportation network there can be exponential
number candidate paths. Therefore, we suggest to follow the second approach of
determining the critical time points at run-time. In this approach only a small
fraction of candidate paths and their cost functions are actually considered while
computation. Now, we formal define critical-time-points and describe our method
of determining them efficiently.

Critical Time Point *A departure-time instant when the earliest arrival path
between a source and destination may change.*

Consider an instance of the ALSP problem on the digraph shown in Fig. 5.4,
where the source is node A, the destination is C, and $\lambda = [0\ 1\ 2\ 3\ 4]$. Here, the
departure-time of $t = 2$ is a critical time point because the earliest arrival path
between node A and C changes for departure-times greater that $t = 2$. Similarly,
$t = 2$ is also a critical time point for the temporal digraph shown in Fig. 5.3a, where
the source node is A and the destination node is D (see Fig. 5.3b). Now, in order
to determine these departure-times, we need to model the total cost of the path.
We use a weight function to capture the total cost of a path. This approach, which
associates a weight function to a path, yields a *path-function* which represents the
earliest arrival time at the end node of the path. A formal definition of the path-
function is given below.

Path Function *A path function represents the earliest arrival time at the end node
of a path as a function of time. This is represented as a time series. A path function
is determined by composing the arrival time functions of its component edges in a
Lagrangian fashion.*

For example, consider the path <A,B,C> in Fig. 5.4. This path contains two
edges, (A, B) and (B, C). The arrival-time function of edge (A, B) is [3 4 5 6 7],

while that of edge (B, C) is [1 2 3 4 6 8 10 12]. Now, the path function of <A,B,C> for start times [0, 4] is determined as follows. As per the given function, if we start at node A at $t = 0$, the arrival time at node B is 3. Now, arrival time at node C through edge (B, C) is considered for time $t = 3$ (Lagrangian fashion), which is 4. Thus, the value of the path function of <A,B,C> for start time $t = 0$ is 4. The value of the path-function for all the start times is computed in similar fashion. This would give the path function of <A,B,C> as [4 6 8 10 12]. This means that if we start at node A at times $t = 0, 1, 2, 3, 4$ then we will arrive at end node C at times $t = 4, 6, 8, 10, 12$. Similarly, the path function of path <A,C> is [5 6 7 8 9] (since it contains only one edge). By comparing the two path-functions we can see that path <A,B,C> has an earlier arrival time (at destination) for start times $t = 0$ and $t = 1$ (ties are broken arbitrarily). However, path <A,C> is shorter for start time $t \geq 2$. Thus, start time $t = 2$ becomes a critical time point. In general, the critical time points are determined by computing the intersection point (with respect to time coordinate) between path functions. In this case, the intersection point between path functions <A,C> and <A,B,C> is at time $t = 2$. Computing these intersection points efficiently is the basis of the CTAS algorithm which is described next.

5.3 Critical Time-Point Based ALSP Solver (CTAS)

The basic computational unit of CTAS computes an earliest arrival path, between the source and destination, for one departure-time and then forecasts a lower bound on the departure-time when the current solution may change. The algorithm restarts its basic computational unit at this forecasted departure-time. This process repeats while there is still a departure-time (in the given interval λ) for which the algorithm does not have a earliest arrival path. At each stage the basic computational unit computes, what we refer to as, a *Sub-Interval Optimal Lagrangian Path*.

Sub-Interval Optimal Lagrangian Path *is a Lagrangian path, P_i, and its corresponding set of time instants ω_i, where $\omega_i \in \lambda$. P_i is the optimal earliest arrival path between the source and the destination for all the departure-times $t \in \omega_i$.*

The Lagrangian path <A,C,D> shown in Fig. 5.3b is an example of sub-interval Optimal Lagrangian Path and its corresponding set $\omega = 0, 1$.

5.3.1 CTAS Algorithm

The algorithm starts by computing the earliest arrival path for the first departure-time in the set λ. Due to non-stationarity, the choice of the path to expand at each step made while computing the optimal path for one particular departure-time may not be valid for later time instants. Therefore, the algorithm stores all the critical time points (potential) observed while computing the earliest arrival path for

---

Algorithm 5 CTAS algorithm

1: Initialize the path intersection table with the first departure-time in λ
2: **while** A earliest arrival path for each time $t \in \lambda$ is to be determined **do**
3:     $t_{cur} \leftarrow$ Min{potential critical-time-points in path intersection table}.
4:     Clear the path intersection table.
5:     Initialize a priority queue with the path functions corresponding to the source node and its neighbors. These path functions are ordered on their value at time t_{cur}.
6:     **while** Destination node is not closed **do**
7:         Path $P_{min} \leftarrow$ ExtractMin.
8:         Add the tail node of P_{min} to the list of closed nodes.
9:         Determine the earliest time when P_{min} would not be top of the queue.
10:        Save that time in the path intersection table as a potential critical-time-point.
11:        Delete all the paths ending on the tail node of P_{min} from the priority queue.
12:        Determine the path functions resulting from the expansion of P_{min}.
13:        Push the newly determined path functions into the priority queue.
14:    **end while**
15:    Output the path where destination node was close as the sub-interval optimal Lagrangian path. Its ω would be interval $[t_{cur} \; Min\{potentialcritical-time-pointsinpathintersectiontable\}]$
16: **end while**

---

one departure-time in a data structure called *path-intersection table*. As discussed previously, a critical time point is determined by computing the time coordinate when the path functions of the candidate paths change their relative order.

Once we have full source to destination path, we look up the potential critical time points stored in the path intersection table. The earliest of these time points represents the first time instant when the current source-destination path may no longer remain optimal. This becomes the forecasted critical-time-point and the re-computation starts from this time instant. Since the path functions store the *arrival time at the tail node* of the path as a function of *departure-times at the source*, the intersection points would automatically represent the departure-time (at the source) when there is a change in relative ordering of the candidate paths.

A pseudocode of the CTAS algorithm is presented in Algorithm 5. The outer loop of ALSP ensures that an earliest arrival path is computed for each departure-time given in the set λ. The inner loop computes a single sub-interval optimal Lagrangian path. Over the course of execution, the algorithm may generate several sub-interval optimal Lagrangian paths for the set λ. Note that it is possible that the two successive sub-interval optimal Lagrangian paths computed by the algorithm may actually have the same physical path. This may happen when the forecasted critical-time-point may turn out to be a false positive.

In each instance of the inner loop (line number 6–14 in Algorithm 5), the algorithm computes a sub-interval optimal Lagrangian path for departure-times greater than (or equal to) the time t_{cur}. Before entering the inner loop, t_{cur} is set to minimum of the potential critical-time-points stored in the path intersection table. Then, a priority queue is initialized with the path functions corresponding to the neighbors of the source node. This priority queue is ordered on the values of the path functions for time t_{cur}. Note that for the first instance of the inner loop, t_{cur} would get set to the first departure-time in set λ.

Inside the inner loop, in each iteration, the algorithm extracts the path (P_{min}) which has the minimum weight for time $t = t_{cur}$ (see line number 7 in Algorithm 5). Following this, the tail node of P_{min} is deemed to be closed for the time coordinate on which the priority queue is ordered (which is t_{cur}). For example, if a path $s-x-y$ was the result of ExtractMin when the priority queue was ordered on time t', then node y is closed for departure-time $t = t'$. This means that one cannot reach node y earlier using any other path (other than $s-x-y$), if they have to depart from node s at time $t = t'$.

After adding the tail node of P_{min} to the closed list, the algorithm computes the intersecting points among the path functions available in the priority queue. The earliest intersection point (considering intersection points in the increasing order of their time coordinate) involving the path function of P_{min} is the earliest time instant when the optimality of this partial path (and its successors) may change. This time instant is saved in the path intersection table as a potential critical-time-point. Following this, the algorithm deletes all paths which end on the tail node of P_{min} from the queue. In the next step, P_{min} is expanded and path functions to all neighbors of the tail node of P_{min} are computed and added to the priority queue. The inner loop terminates when the destination node is closed. Following this, the sub-interval optimal Lagrangian path is output. Its ω would be interval $[t_{cur}$ $Min\{potentialcritical - time - pointsinpathintersectiontable\}]$

In the next iteration of the outer loop, computation starts from the earliest of the potential critical time points stored in the path intersection table. In a worst case scenario, this value could just be the next departure-time instant in the set λ (one more than previous t_{min}). In such a case, the earliest arrival path determined by the previous instance of the inner loop was optimal for only one departure-time. The algorithm terminates when we have an optimal path for all departure-times in the set λ.

5.3.2 Execution Trace

Figure 5.5 gives an execution trace of the CTAS algorithm on a sample ALSP problem instance. The top of the figure illustrates the input temporally roadmap modeled as a temporal digraph. In this ALSP problem instance, S is the source node, D is the destination node, and desired set of departure-times $\lambda = 0, 1, 2, 3$. With the intention of keeping the discussion concise, we provide a trace for computing only the first of the sub-interval optimal Lagrangian paths for the given ALSP problem instance.

As described earlier, the inner loop of the algorithm builds a single sub-interval optimal Lagrangian path whose optimality starts from the earliest of the potential critical time points. For the first iteration, this would be $t = 0$ (the first start time instant in λ), thus, $t_{cur} = 0$. Trivially, the source node is added into the list of closed nodes for time $t = 0$. Path functions for its immediate neighbors (path <S,B> and <S,C>) are computed and added them to the priority queue (see Step 1 in Fig. 5.5).

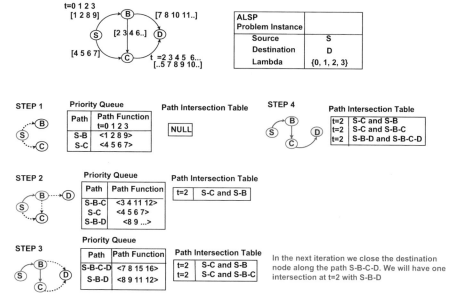

Fig. 5.5 Execution trace of CTAS algorithm

In this case, the path functions just happen to be their edge weight functions. The priority queue is ordered on the value of path functions at t_{cur} (which is 0 in this instance of the inner loop).

In the first iteration of the inner loop, the algorithm chooses the path whose path function has minimum weight at t_{cur}. At this point the priority queue has two path functions, $<S,B>$ and $<S,C>$ (see Fig. 5.5). The algorithm chooses $<S,B>$ because it has lowest cost for t_{cur}. Node B is added to the list of closed nodes. Following this, the algorithm stores the earliest of the intersection points between the path function of $<S,B>$ and any other path function in the priority queue, which happens to be with $<S,C>$ at $t = 2$. This information is stored in the path intersection table as a potential critical time point. Following this any path which ends on node B other than $<S,B>$ is deleted from the priority queue. $<S,B>$ is now expanded and path functions for its neighbors—$<S,B,C>$ and $<S,B,D>$—are computed and added to the priority queue (see Step 2 in Fig. 5.5).

In the next iteration of the inner loop, $<S,B,C>$ would be the result of ExtractMin. Node C would be added to list of closed nodes. Path function of $<S,B,C>$ intersects with the path function of $<S,C>$ at time $t = 2$. This information is added into the path intersection table as potential critical time point. Following this, the algorithm would delete paths ending on node C (other than $<S,B,C>$) from the queue. As a result path $<S,C>$ is deleted from the queue.

At this stage, the priority queue has two path functions, $<S,B,C,D>$ and $<S,B,D>$ (see Step 3 in Fig. 5.5). ExtractMin would now give us $<S,B,C,D>$. The earliest intersection point between the path function of $<S,B,C,D>$ and that of <S,B,D> is computed, which happens to be at $t = 2$. This is added to the path intersection table. <S,B,C,D> is a complete source to destination path. This is the terminating condition of the inner loop. After coming out of the inner loop, the algorithm outputs <S,B,C,D> as the sub-interval optimal Lagrangian path with $\omega = \{0, 1, \min\{potentialcriticaltimepoints\}\}$.

In the next instance of the inner loop, the algorithm builds the sub-interval optimal Lagrangian path starting at the earliest of the potential critical time points stored in the path intersection table. This happens to be 2 in our example. Thus, the next iteration of the inner loop works with $t_{cur} = 2$. One may notice that the algorithm did not compute the shortest path for start time $t = 1$. The fact that the next iteration of the CTAS algorithm starts with $t = 2$ shows that it saves computation. The algorithm would terminate when we have earliest arrival paths for all the time instants in the set λ.

5.4   Correctness and Completeness CTAS Algorithm

The CTAS algorithm divides a given departure-time interval into a set of disjoint sub-intervals. Within these intervals, the optimal path does not change. The correctness of the CTAS algorithm requires the path returned for a particular sub-interval to be optimal and the completeness of the algorithm guarantees that none of the critical time points are missed. The correctness of the CTAS can be easily argued on the basis greedy nature of the algorithm. Lemma 5.1 shows that the CTAS algorithm does not miss any critical time point. Using Lemma 5.1, Theorem 5.1 proves the completeness of the CTAS algorithm.

Lemma 5.1 *The CTAS algorithm recomputes the optimal path for all those departure-times $t_i \in \lambda$, where the optimal path for previous departure-time t_{i-1} can be different from the optimal path for departure-time t_i.*

Proof Consider the sample network shown in Fig. 5.6, where an earliest arrival-time path has to be determined between node s and d for all departure-times in the discrete time interval $\lambda = [1, 2 \ldots T]$. First, the source node is expanded for the time instant $t = 1$. As a result, path functions for all the neighbors $< s, 2 >$, $< s, 1 >$,

Fig. 5.6 Network for
Lemma 5.1

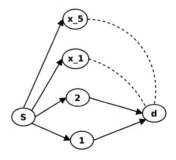

$< s, x_i >$ are added to the priority queue. With loss of generality assume that path $<s,1>$ is chosen in the next iteration of the inner loop. Also assume that the earliest intersection point between path functions $<s,x_i>$ and $<s,1>$ is at $t = \alpha$ between $<s,1>$ and $<s,2>$. Now, the queue contain paths $< s, 1, d >, < s, 2 >, < s, x_i >$. Here we have two cases. First, path $< s, 1, d >$ has lower cost for departure-time $t = 1$. Second, path $<s,2>$ has lower cost for departure-time $t = 1$.

Considering the first case, without loss of generality assume that the earliest intersection point between the path functions $<s,1,d>$ and $<s,2>$ is at $t = \beta$. Note that both $t = \alpha$ and $t = \beta$ denote the departure times at the source node. Consider the case when $\beta \leq \alpha$ (Note that β cannot be greater than α as all the edges have positive weights). In such a case the optimal path is recomputed for time $t = \beta$ and the path $<s,1,d>$ is closed for all departure-times $1 \leq t < \beta$. Assume for the sake of contradiction, that there is a path P_x from source to destination that is different from path $<s,1,d>$ which is optimal for departure time $t_x \in [1, \beta)$. Assume that $P_x =< s, x_1, x_2, x_3, \ldots, d >$. This means that path $< s, x_1 >$ had least for time $t = t_x$. However, by the nature of the algorithm, this path would have been expanded instead of path $< s, 1 >$ (a contradiction). Moreover, as all the travel-times positive, if sub path $<s,x_1>$ was not shorter than $<s,1,d>$ for departure-times earlier than $t = \beta$. Any positive weight addition to the path function (through other edges) cannot make P_x shorter than $<s,1,d>$.

Consider the second case when $<s,2>$ had lower cost for start time $t = 1$. Now, path $<s,2>$ would be expanded and path function $<s,2,d>$ would be added to priority queue. Again, assume that path functions $<s,1,d>$ and $<s,2,d>$ intersect at time $t = \beta$. A similar argument can be given for this case as well.

Corollary 5.1 *(Corollary to Lemma 5.1) Given a instance of CTAS which started from node S at time t_d and terminated when node D was closed. Now, if the earliest forecasted critical time point was t_{ctp}^{min}, then shortest path determined between S and D is optimal from all departure-times t_d through $t_{ctp}^{min} - 1$.*

Theorem 5.1 *CTAS algorithm is complete.*

Proof There may be several sub-interval optimal Lagrangian paths P_i over set λ. Each P_i is associated with a set of time instants ω_i, where $\bigcup_{\forall i \in |P_{sd}|} \omega_i = \lambda$. The completeness proof of the CTAS algorithm is presented in two parts. First, using Lemma 5.1 we can conclude that the CTAS algorithm does not miss any departure time instant when the shortest path changes. Secondly, the outer loop iterates until the algorithm determines a shortest path for all the time instants in set λ. This happens when the earliest of the forecasted CTPs t_{min} falls outside λ. This proves the completeness of the algorithm.

Discussion The CTAS algorithm shows better performance than a naive approach which, determines the shortest paths for each departure-time in the user specified time interval. However, this happens only when the ratio of the number of critical time points to that of departure-times instants is low. This ratio can be denoted as the change probability shown by Eq. (5.1).

$$change\_probability = \frac{\#critical\ time\ points}{\#departure - times} \tag{5.1}$$

When the change probability is nearly 1, there would be a different optimal path for each departure-time. In this worst case scenario, the CTAS approach would also have to recompute the optimal path for each departure-time.

5.5 Experimental Evaluation of CTAS

Experiments were conducted to evaluate the performance of CTAS algorithm as different parameters were varied. The experiments were carried on a real dataset containing the highway road network of Hennepin county, Minnesota, provided by NAVTEQ [54]. The dataset contained 1417 nodes and 3754 edges. The data set also contained travel times for each edge at time quanta of 15 min. Figure 5.3 shows the speed profiles for a particular highway segment in the dataset over a period of 30 days. As can be seen, the speed varies with the time of day. For experimental purposes, the travel times were converted into time quanta of 1 min. This was done by replicating the data inside time interval. The experiments were conducted on an Intel Quad core Linux workstation with 2.83 Ghz CPU and 4 GB RAM. The performance of CTAS algorithm was compared against a modified version of the existing BEST start time algorithm proposed in [23].

Modified-BEST (MBEST) Algorithm The MBEST algorithm consists of two main parts. First, the shortest path between source and destination is determined for all the desired start time instants. Second, the computed shortest paths are post-processed and a set of distinct paths is returned. The MBEST algorithm uses a label correcting approach similar to that of BEST algorithm, proposed in [23] to compute the shortest paths between source and destination. The algorithm associates two lists viz, the arrival time list and ancestor list, with each node. The arrival time array, $C_v[t]$, represents the earliest arrival time at node v for the start time t at the source. The ancestor array, $An_v[t]$, represents the previous node in the path from source for time t. These lists are updated using Eq. (5.2), where γ_{uv} represents the earliest arrival time series of the edge (u, v). The algorithm terminates when there are no more changes in the arrival time list of any node.

$$C_v[t] = \min\{C_v[t], \gamma_{uv}[C_u[t]]\}, uv \in E \tag{5.2}$$

Experimental Setup The experimental setup is shown in Fig. 5.7. The first step of the experimental evaluation involved combining the travel time information along with the spatial road network to represent the ST network as a Time-aggregated graph. A set of different queries (each with different parameters) was run on the CTAS and the MBEST algorithms. The following parameters were varied in the experiments: (a) length of the time interval over which shortest paths were desired

Fig. 5.7 Experimental setup

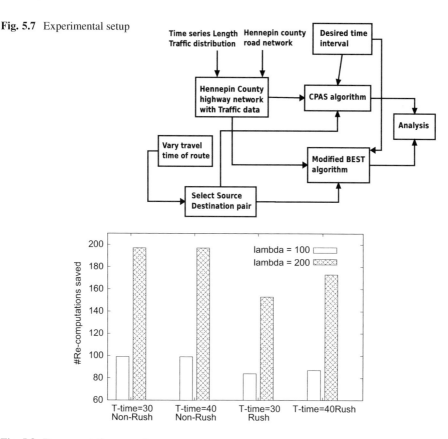

Fig. 5.8 Re-computations saved

($|\lambda|$), (b) total travel time of the route, (c) time of day (rush hour vs non-rush hours). A speedup ratio, given by Eq. (5.3), was computed for each run. The total number of re-computations avoided by CTAS was also recorded for each run. In worst case, the shortest paths may have to be re-computed for each time instant in the interval λ. The total number of re-computations saved by CTAS is the difference between $|\lambda|$ and re-computations performed.

$$speed - up\ ratio = \frac{MBEST\ runtime}{CTAS\ runtime} \tag{5.3}$$

Number of Re-computations Saved in CTAS Figure 5.8 shows the number of re-computations saved by the CTAS algorithm. The experiments showed that more saving was gained where paths were shorter. Similarly, fewer number of re-computations were performed in case of Non-rush hours. This is because there were fewer intersections among the path-functions.

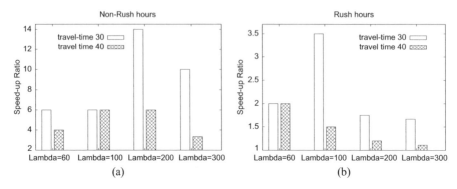

Fig. 5.9 Effect of lambda on speed-up

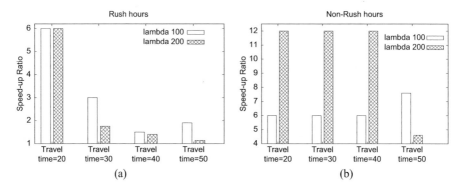

Fig. 5.10 Effect of travel time on speed-up

Effect of Length of Start Time Interval ($|\lambda|$) This experiment was performed to evaluate the effect of length of start time interval (λ) over which the shortest path was desired. Figure 5.9a shows the speed-up ratio for Non-rush hours and Fig. 5.9b shows the speed-up ratio for Rush hours. The speed ratio was calculated for paths with travel time 30 and travel time 40. These travel times indicate the time required to travel on these paths during non-rush hours when there is no traffic. The experiments showed that run-time of both CTAS and MBEST increased with increase in the lambda. Runtime of MBEST increases steadily whereas CTAS increases very slowly with *lambda* for a travel time of 30. However, the run-time of CTAS increases rapidly for travel time of 40.

Effect of Total Travel Time of a Path This experiment was performed to evaluate the effect of total travel time of a path on the candidate algorithms. Figure 5.10 shows the speed-up ratio as the total travel time of the path was varied. The experiments showed that the runtime of CTAS algorithm increased with a corresponding increase in the total travel time of the path, whereas the runtime of the MBEST algorithm remained the same. This is because, CTAS algorithm follows a Dijkstra's like approach and expands the paths, but MBEST is follows a label correcting approach and terminates when there no more changes in the arrival time array of any node.

Effect of Different Start Times Experiments showed that better speed-up was obtained for Non-rush hours than the rush hours (see Figs. 5.10 and 5.9). This is because there are fewer number of intersection points in case of non-Rush hours.

5.6 Conclusions

The All-start-time shortest Lagrangian shortest path problem (ALSP) is a key component of applications in transportation networks. ALSP is a challenging problem due to the non-stationarity of the spatio-temporal network. Traditional A* and Dijkstra's based approaches incur redundant computation across the time instants sharing a common solution. The proposed Critical Time-point based ALSP Solver (CTAS), reduces this redundant re computation by determining the time points when the ranking among the alternate paths between the source and destination change. Theoretical and experimental analysis show that this approach is more efficient than naive particularly in case of few critical time points.

Chapter 6
Advanced Concepts: Bi-Directional Search for Temporal Digraphs

Recall the all-start-time Lagrangian shortest path (ALSP) problem defined in Chap. 5. In this chapter, we would introduce a bi-directional search for solving this problem. Unlike the bi-directional search discussed in Sect. 4.4 of Chap. 4, here, the backward search would happen on the reverse of the input temporal digraph itself. To prevent ambiguity of terms, we would refer to the backward search, in this algorithm, as *trace* search. Also, we would be using the phrase *temporal bi-directional* search to distinguish from the traditional bi-directional search algorithms which use a reverse of the lower bound graph for their backward searches (as discussed in Sect. 4.4 of Chap. 4). As a quick recollection, following is a formal definition of the ALSP problem, just the assumptions are changed for the temporal bi-directional search.

Inputs

1. A temporally detailed roadmap dataset represented as a temporal digraph $G = (V, E)$, where V is the set of vertices, and E is the set of directed edges.
2. Each edge $e \in E$ is associated with a cost function δ, which is discrete in nature. This function gives the arrival time (or the earliest arrival time in case of non-FIFO; refer Sect. 4.1 in Chap. 4) at the end node of the edge for different departure-times at the starting node of the edge.
3. A source s and a destination d pair where $\{s, d\} \in V$.
4. A discrete time interval λ over which the fastest path between s and d needs to be determined.

Output The output is a set of routes, P_{sd}, from s to d where each route $P_i \in P_{sd}$ is associated with a set of departure time instants ω_i, where ω_i is a subset of λ.

Objective Function Each path in P_{sd} is the earliest arrival path between s and d for its respective time instants ω_i.

Assumptions We assume that the length of the time horizon over which the temporal digraph is considered is finite. The cost function δ is a discrete time series.

© Springer International Publishing AG 2017
V.M.V. Gunturi, S. Shekhar, *Spatio-Temporal Graph Data Analytics*,
https://doi.org/10.1007/978-3-319-67771-2_6

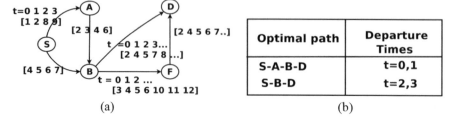

Fig. 6.1 Sample input and output of ALSP problem. (a) Sample input temporal digraph. (b) Output of ALSP problem

Also, we assume that the edge cost function δ follows, what we refer to as, *strong fifo property*. This means that an earlier departure on any route strictly implies an earlier arrival at its terminal node. This assumption is needed while proving the correctness of the algorithm.

Sample Problem Instance Consider the sample temporally detailed roadmap shown in Fig. 6.1. The edge cost function δ is representing the arrival time series of the edge. Note that in case of strong FIFO, earliest arrival time series would same as the arrival time series. An instance of the ALSP problem on this input, with source S destination D and $\lambda = [0, 1, 2, 3]$, is shown in Fig. 6.1b. Here, the path S-A-B-D is optimal for departure-times $t = 0$ and $t = 1$, whereas the path S-B-D is optimal for departure-time $t = 2$ and $t = 3$.

Outline of Rest the Chapter A description of the computational structure of the temporal bi-directional search is presented in Sect. 6.1. This section introduces the concepts of *forward* and *trace* search, which later form the building blocks of a temporal bi-directional search for the ALSP problem. Section 6.2 presents the novel *impromptu rendezvous* [29] termination condition for a temporal bi-directional search. In Sect. 6.2.1, we present the actual algorithm.

6.1   Computational Structure of Temporal Bi-Directional Search

Recall that the basic computational unit in a critical time point (CTP) based approach was to compute a single *sub-interval optimal Lagrangian path*. By computing successive sub-interval optimal Lagrangian paths, a CTP based approach was able to compute the complete solution of a given ALSP instance. In a temporal bi-directional search, we would use both forward search (starting from source) and trace search (starting from the destination) to determine a single sub-interval optimal Lagrangian path. This is different from the approach discussed in Chap. 5 which was based on forward search alone. We now present the core ideas of a forward and a trace search on temporal digraphs. In the next section, we would put these ideas together along with a novel termination condition to determine a single sub-interval optimal Lagrangian path.

6.1.1 Forward Search Basic Concepts

The goal of the forward search is explore candidate paths from the source node for a departure-time in given interval of departure-times (λ). The search terminates (for a particular departure-time) when the destination node is closed. The forward search, while computing shortest path for a departure-time, maintains some information to forecast the critical time point (forward critical time point) at which the search needs to re-compute the shortest path.

Forward Critical Time Point *A departure time instant at a source (as determined by the forward search) when the shortest path between a source and destination changes.*

Consider an instance of a forward search on the temporal digraph shown in Fig. 6.2, where the source is node S, the destination is D, and departure-times $t = 0, 1, 2, 3$. Here, departure time $t = 2$ is a forward critical time point because the shortest path between nodes S and D changes for departure times greater that $t = 2$. In order to determine these critical time points, we need to model the total cost of the path. This paper proposes using a weight function to capture the total cost of a path. This approach yields a *forward path-function*, which represents the earliest arrival time at the end node of the path as a function of time. A formal definition of the path-function is given below.

Forward Path Function *A forward path function represents the arrival time at the end node of a path as a function of time. This is represented as a time series. A path function is determined by composing the arrival times on its component edges in a Lagrangian fashion.*

For example, consider the path *S-A-D* in Fig. 6.2. This path contains two edges, (S, A) and (A, D). The arrival time series of edge (S, A) is [1 2 3 4], while that of edge (A, D) is [1 2 3 6 7 8]. The forward path function of *S-A-D* for departure times [0, 1, 2, 3] is determined as follows. A journey starting at node S at $t = 0$ would arrive at node A by $t = 1$. Now, arrival time at node D through edge (A, D)

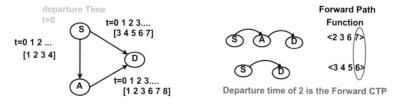

Fig. 6.2 Forward critical-time points. δ of edges contains earliest arrival time values

is considered for time $t = 1$ (Lagrangian fashion), which is 2. Thus, the value of the forward path function of S-A-D for departure time $t = 0$ is 2. The value of the path-function for all the departure times can be computed in similar fashion. This would give the forward path function of S-A-D as [2 3 6 7] (shown in Fig. 6.2). This means that if we depart from node S at times $t = 0, 1, 2, 3$ then we will arrive at node D at times $t = 2, 3, 6, 7$. Similarly, the forward path function of path S-D is [3 4 5 6] (since the path contains only one edge). By comparing the two path-functions we can see that path S-A-D has an earlier destination arrival time for times $t = 0$ and $t = 1$. However, path S-D is shorter for departure times $t \geq 2$. Thus, departure time $t = 2$ becomes a forward critical time point. In general, critical time points are determined by computing the intersection point (with respect to time coordinate) between candidate path functions. In this case, the path functions for S-A-D and S-D intersect at time $t = 2$.

Forward Search Algorithm   Uses an enumerate and prune style similar to Dijkstra's. A brief pseudocode is given in Algorithm 6. The algorithm maintains a priority queue PQ_{for} containing the forward path functions ordered based on the current departure-time, t_{d_i}, under consideration. In each iteration the algorithm selects the path with the min weight (FP^{min} in Algorithm 6). Then, it computes the earliest intersection point of FP^{min} with other path functions in PQ. This intersection point is stored in the path intersection table (referred to as $Pinter_{for}$). Following this it expands the corresponding path of FP^{min} to create new forward path functions for the neighbors of its tail node. The newly created path functions are pushed into PQ_{for}. Also, at this stage we add the tail node of FP^{min} to the list of closed nodes and all path functions ending on that node are deleted from PQ_{for}. This process continues until the destination node is closed. At termination, the earliest of the time points stored in $Pinter_{for}$ ($min\{Pinter_{for}\}$) then becomes the forecasted *forward critical time point*. It is important to note that a single run of the algorithm in Algorithm 6 gives a single *sub-interval lagrangian path* between S and D with $\omega = [t_{d_i} \ldots (min\{Pinter_{for}\} - 1)]$. Thus, a repeated application of this algorithm, until all departure-times in λ are covered, would give us a solution for the ALSP problem. Our previous approach discussed in Chap. 5 follows this exact procedure.

Algorithm 6 Forward search algorithm at time τ

1: Initialize priority queue $PQ_{for} \leftarrow$ path functions of S and its neighbors. Path functions in PQ_{for} are ordered for their value at τ.
2: **while** D is not expanded **do**
3:     $FP^{min} \leftarrow$ min weight path function from PQ_{for}
4:     $ITP \leftarrow$ earliest intersection point of FP^{min}
5:     Save ITP in $Pinter_{for}$
6:     Expand FP^{min} to its neighbors
7:     Push the newly determined path functions into PQ_{for}
8:     Add tail node of FP^{min} to list of closed nodes
9:     Delete paths ending on tail of FP^{min} from PQ
10: **end while**
11: forecasted forward CTP $\leftarrow min\{Pinter_{for}\}$

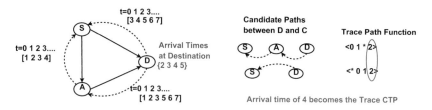

Fig. 6.3 Illustration of Trace Critical-time points. δ of edges contains earliest arrival time values

6.1.2 Trace Search Basic Concepts

The goal of a trace search is to compute the fastest paths from other nodes to a destination such that these paths arrive (at the destination) at a particular time instant within an estimated arrival time interval (λ_{rev}). Similar to the forward search, a trace search maintains some additional information to forecast the time point (trace critical time point) when a certain shortest path to the destination changes. However, in this case, candidate paths are explored in a *reverse temporal digraph* (STN_{rev}). This network is defined on the same set of nodes V as the original temporal digraph, but contains a flipped edge (v, u) for each (u, v) in the original network.

Trace Critical-Time Point *An arrival time instant at a destination (as determined by the trace search) when the shortest path to the destination from a source changes.*

Consider an instance of the trace search on the temporal digraph shown in Fig. 6.3, where the destination is D, and $\lambda_{rev} = [2\ 3\ 4\ 5]$. This means we want to find the shortest paths from all other nodes to node D such that they arrive at D at times $t = 2, 3, 4, 5$. The trace search is performed on a reverse ST network where each edge in the original ST network is reversed (showed as dotted edges on left of Fig. 6.3). The arrival times series in the original network are interpreted as an *inverse-mapping* from arrival times to departure-times. For instance, in Fig. 6.3, consider the original edge (S,D) (shown in bold) and its corresponding reverse (shown using a dotted arrow). The arrival time series of the original edge was [3 4 5 6 7], which means that a departure at $t = 0$ at node S would have an arrival at D at $t = 3$. Since this is a trace search, we would read this time series as: *If need to arrive at D at $t = 3$, what is the latest departure time at S?*.

In our previous trace search instance with destination D and $\lambda_{rev} = [2\ 3\ 4\ 5]$ in Fig. 6.3, the arrival time $t = 4$ is a trace critical time point because the shortest path to D from S changes for arrival times greater that $t = 4$. Again, in order to determine these arrival time instants, we model the total cost of the path using a weight function. This weight function, which we call *trace path function*, is defined formally below.

A **Trace Path Function** *represents the latest departure times at the end node as a time series. Each item in this time series represents the latest departure at the end node for the corresponding arrival time at the start node of the path.* Similar to a forward path function, a trace path function is determined by combining the arrival times on its component edges in a Lagrangian fashion. The only difference is that, due to inverse mapping, we look backwards while composing the time series.

For instance shown in Fig. 6.3, we have two paths to D. Path D-A-S is composed of edges (D, A) and (A, S) (the original edges in reverse). In order to reach D at time $t = 2$ (first arrival time in λ_{rev}), we need to start at A by $t = 1$ at the latest. Similarly, in order to reach A at $t = 1$, we need to start at S by $t = 0$ at the latest. Thus, the first value in the trace path function of D-A-S would be 0. Values for other arrival times in $\lambda_{rev} = [2\ 3\ 4\ 5]$ can be computed in a similar fashion. While composing the arrival time series for the component edges, we consider only the latest departure time that incurs no waiting. Also, in some cases, the arrival times cannot be mapped backwards. For instance, in our previous example of $\lambda_{rev} = [2\ 3\ 4\ 5]$, the arrival time 4 cannot be mapped back along edge (D, A). In other words, there does not exist a departure-time at A such that its corresponding arrival time (with no waiting) at D is 4. Note that a departure-time of 2 at A would not be considered as it would arrive at D early and wait. Cases like these are represented using '*' (denoting $-\infty$) in the trace function.

The above process gives the following trace function of D-A-S: <0 1 * 2>. This means, in order to reach D at times 2,3 and 5 we need to start from S (along S-A-D) by times 0,1, and 2 (at the latest). We use $-\infty$ as the numeric value for '*' as it makes it easier to compute the trace critical time points during the course of the algorithm. Figure 6.3 illustrates the trace functions of the two candidate paths D-A-S and D-S over the arrival times (at D) $\lambda_{rev} = [2\ 3\ 4\ 5]$. We can compare these trace functions to determine the trace critical time points. Notice that a larger trace path function value implies the path is shorter. For instance, the trace function value of D-A-S for $t = 3$ (second arrival time in λ_{rev}) is 1, whereas the same for D-S is 0. This means D-A-S is the shorter path (for latest departure) to D for the arrival time $t = 3$. In Fig. 6.3 such comparison gives arrival time 4 as the trace critical time point. In other words, we should use path S-A-D (for latest departure) in-order to arrive at D at times 2, 3, and path S-D (for latest departure) for others.

We could construct an algorithm similar Algorithm 6 using the concepts of trace critical-time-point and trace path function. Given a source, a destination, and a desired destination arrival time interval λ_{rev}, this algorithm would start exploring paths from destination on a reverse temporal digraph for a particular arrival time t_{a_i} in λ_{rev} and make a conservative forecast of trace critical-time point. The priority queue in the trace search would be sorted in ascending order on the difference between arrival-time at destination and the values in the trace function. This type of ordering is followed to ensure that the node closest to the destination is closed first, an essential requirement for correctness. Correctness claim of trace search is given in Lemma 6.1 (Sect. 6.3). This is necessary (as shown later) while establishing the correctness of the Bi-directional CTAS algorithm in Sect. 6.

We now begin our discussion on the temporal bi-directional search for the ALSP problem. During the discussion, we may sometimes drop the prefix "trace" or "forward" from critical-time-points and path functions when the context is clear. Also critical-time-points may be abbreviated as CTP for convenience.

6.2 Temporal Bi-Directional Search for ALSP Problem

A temporal bi-directional search for ALSP problem mainly consists of a forward and a trace search. The forward search starts the search from the source with the given set of departure-times in λ. The trace search starts from the destination node and explores temporal digraph in reverse (network with edges reversed) using the trace path functions. Designing an efficient bi-directional search for the ALSP problem poses two challenges. First, unlike the forward search which explores over a given set of departure-times, appropriate "departure-times" for the trace search at the destination are not known in advance. Second, designing a suitable termination condition which gives good performance while ensuring correctness is non-trivial.

"Departure-Times" for Trace Search To address the first challenge, we compute a bound on the arrival times at the destination. These bounds are computed by executing a single source shortest path algorithm developed for temporal digraphs [23] for the first and the last departure-time in λ. Given the fifo nature of our network, we can claim that arrival times for all the other departure-times in λ would fall inside the interval defined by the arrival time for first and last departure-times. The range defined by the arrival times at the destination (for first and last departure-time) form the set of "departure-times" (denoted by λ_{rev}) for the trace search. To avoid ambiguity, we use the term *trace-times* instead of "departure-times" when discussing the trace search.[1] We use the term *departure-times* in context of the forward search or the problem instance (λ).

Figure 6.4 shows the upper and lower bounds defining the range λ_{rev} for our ALSP instance in Fig. 6.1. It is important to note that although these bounds are tight, there may not always be a bijective mapping between the departure-times and the arrival times. For instance, the four departure-times ($\lambda = \{0, 1, 2, 3\}$) in our earlier instance of ALSP (Fig. 6.1) were mapped to a broader range of five arrival times ($\lambda_{rev} = \{7, 8, 9, 10, 11\}$) in Fig. 6.4. Such instances happen due to an increase in the cost of the path, making some arrival times at the destination infeasible.

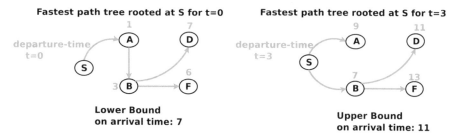

Fastest path tree rooted at S for t=0   Fastest path tree rooted at S for t=3

Lower Bound on arrival time: 7   **Upper Bound on arrival time: 11**

Fig. 6.4 Bounds on arrival time

---

[1]Trace-times refer to time points in λ_{rev}, while departure-times refer to time points in λ (in the ALSP instance).

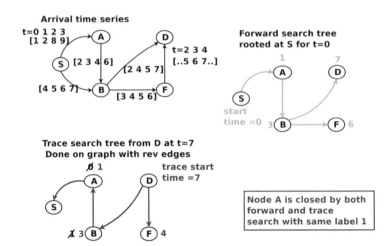

Fig. 6.5 Illustration of impromptu rendezvous condition. Temporal Digraph from Fig. 6.1a has been replicated here for convenience. δ of edges represent earliest arrival times

***Impromptu rendezvous* Condition** Second challenge for bi-directional search was designing an appropriate termination which gives good performance while ensuring correctness. Termination conditions for bi-directional searches work only when the trace search is done on the underlying static graph [12, 15, 51–53] (refer Sect. 4.4 in Chap. 4) cannot be applied in this case. The trace search in these studies is used to prune the search space for the forward search to continue efficiently until destination node is closed. In contrast, trace search in this method is executed on the reverse of the input temporal digraph itself, thus, it allowing us to avoid revisiting the nodes already closed by the trace search. We now describe the *impromptu rendezvous* condition on our previous temporal digraph shown in Fig. 6.1a.

Consider again the temporal digraph illustrated in Fig. 6.1a (replicated in top left of Fig. 6.5). Suppose we want to determine the shortest path between S and D which departs S at time $t = 0$. A bi-directional search for this problem would involve a forward search from node S and a trace search from node D. Clearly, the forward search from S would start at time $t = 0$. By contrast, the choice of trace-time for the trace search is non-trivial and is discussed next with an intuitive explanation of the termination condition.

Figure 6.5 contains a forward search tree (explored until node D was closed) from node S for time $t = 0$ (top right). This can be considered as a instance of forward search algorithm with departure time $t = 0$ run until node D was closed. The figure also contains a trace search tree from node D which starts at time $t = 7$. This can be considered as instance of trace search being executed from D on reverse temporal digraph until node S was closed. The numbers written beside the nodes in the figure represent their respective distance labels when they were closed. For instance, the forward search closed the node F at time 6, i.e, the shortest path to F from S arrives at time 6. On the other hand, the trace search closed node F at time

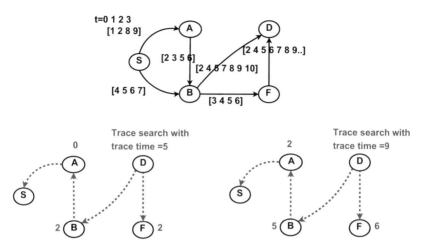

Fig. 6.6 Sample trace search tree on the temporal digraph from Fig. 6.1a for other departure-times

$t = 4$. This means we need to depart node F at the latest by $t = 4$ in order to arrive at D by $t = 7$. A comparison of these two trees shows that nodes A and B happen to have been closed with the same label by both the forward and the trace search. In other words, both the forward (starting at S at $t = 0$) and the trace search (starting at D at $t = 7$) agree on node A and B along the path S-A-B-D, which happens to be the shortest path between S and D for $t = 0$ departure. This shows that when the trace-time of a trace search at the destination is correct (shortest path to D from S indeed arrives at $t = 7$), *both forward and trace searches close the nodes on the shortest path with the same labels.*

If the trace search does not start at the optimal time, there are only other possibilities: it started *earlier* or *later* than the optimal time. Figure 6.6 illustrates these cases. The trace tree at time $t = 5$ (left in the figure) exemplifies the case where the trace search started early. The figure shows that all the nodes in the tree were closed at an earlier time than the forward search tree shown in Fig. 6.5. In other words there was no agreement between the forward and trace search on any node. On the other hand, a trace search starting later than the optimal time (right in the Fig. 6.6) could lead to agreement between the two searches on a node which does not lie on the shortest path. A comparison with the forward search shown in Fig. 6.5 shows that both the searches have agreed on node F, *which does not lie on the shortest path to D.* These observations lead to termination condition we call, the *impromptu rendezvous*, which states: it is sufficient to terminate when both the forward and trace search close a node with the same label, provided the trace-time is not an over-estimate of the optimal arrival time. This is formally stated as Lemma 6.3 in Sect. 6.3.

6.2.1 BD-CTAS Algorithm

In this section a *temporal Bi-Directional* based *Critical-Time-point Alsp Solver* (BD-CTAS) for the ALSP problem. The forward and the trace search along with the impromptu rendezvous condition form the key parts of this algorithm. In the algorithm, the forward and the trace search start from source and destination respectively. The impromptu rendezvous condition is only used as a *sufficiency* condition for terminating the algorithm. It is possible for the algorithm to run without meeting this condition. This happens when the trace-time of the trace search is earlier than the optimal arrival time of the shortest path (Lemma 6.2). In such cases, the algorithm terminates only when the destination node is closed by the forward search (similar to the CTAS algorithm).

Structurally, both the forward and trace search components of this algorithm are similar to the CTAS algorithm. They both explore candidate paths through path functions (or trace functions). Further, whenever a node is closed, the earliest intersection of its forward function (or trace function) is computed with other paths in the priority queue. The priority queue for the forward search is sorted in ascending order on the values in the path functions, while the priority queue in the trace search is sorted (also in ascending order) on the difference between the trace-time at the destination and the values in the trace function. This kind of ordering in the trace search ensures that the closest node (to the destination) is closed first, an essential requirement for correctness. In order to integrate *impromptu rendezvous* termination into the BD-CTAS algorithm, we need to ensure that the trace-time is never an overestimate of the optimal arrival time at the destination (had the forward search been allowed to proceed until the destination).

To this end, the algorithm employs an incremental strategy for determining the next times of the trace and forward search. Given the correctness of the trace-time of a trace search (i.e., it is not an overestimate), a pair of forward and trace searches explore the ST network to determine a single *sub-interval optimal lagrangian path* and determine the time instants for the next pair of searches. We refer to this unit of work as *one iteration* of the BD-CTAS algorithm. Here, the algorithm simply ensures that the next trace-time for the trace search is not an overestimate of the optimal arrival time. This process starts at the first departure-time in λ and continues until shortest paths for all the departure-times given in λ are determined. To ensure the correctness for the first departure-time in λ, the algorithm uses the first time point in the range defined by λ_{rev} (described earlier in the section) as the trace-time of the trace search. Recall that, by construction this trace-time is guaranteed not be an overestimate; in fact, it is the correct destination arrival time. This means that the algorithm does more work for the first departure time in λ. However, this extra cost is usually compensated by the savings generated by the impromptu rendezvous.

We now describe the approach for determining the next times after a pair of forward and trace searches. Assuming that the trace-time of the trace search in the previous instance was not an overestimate, one of the following two cases can be encountered during the execution. Case (a): the algorithm terminated because *impromptu rendezvous* condition was met. Which means both the trace and forward search closed a node with same label. Case (b): the algorithm terminated because the forward search closed the destination node.

Case (a): Impromptu rendezvous Was Satisfied The *impromptu rendezvous* condition guarantees that whenever a node v is closed by both forward and trace search with the same label, the combined path from source to v and v to destination is optimal for its respective departure-time (of the forward search) at source. However, for a more efficient solution for the ALSP problem, we need to forecast a minimum duration for which this path remains optimal. In the CTAS algorithm (Sect. 5.3 in Chap. 5) this was done by choosing the earliest of the intersection points as the forecasted critical-time-point (CTP), the next departure-time for re-computation. In other words, the path was guaranteed to be optimal between the these two departure-times at the source. A natural extension of such an approach to a temporal bi-directional search would be to take the minimum of forward and trace critical time points to determine the next departure and trace times. Note that the trace critical time points would be in terms of arrival times at destination and would generally be higher than the forward CTPs. In order circumvent this, we need to take the minimum over the mutual distance (in time) between CTPs and their "reference time points" (departure-times of forward and trace-time of trace searches).

In order to guarantee the correctness of the above approach in a general case, we need to incorporate the notion of *Lagrangian agreement duration* into the process of finding the minimum of the forward and trace CTPs. Figure 6.7 illustrates this concept with an instance of a temporal bi-directional search on a sample ST network (top left corner of the figure). Here, we want to determine the shortest path between S and D for departure times instants $\lambda = [0\ 1\ 2\ 3]$. The figure shows that path S-B-D is preferable for departure times $t = 0, 1$ (top right corner of the figure). Here, both the searches happen to close node B with the same label. The *Lagrangian agreement duration* is determined by comparing the forward and trace path functions ending at B (shown in Purple in Fig. 6.7). In this case, both path functions agreed for a duration of 2 time units. By this stage, the forward search encountered one forecasted CTP ($t = 3$) due to the intersection between path functions of S-B and S-C (shown using $\{\}$ in figure). Similarly, the trace search encountered a forecasted CTP ($t = 6$) due to the intersection between D-B and D-C (trace functions shown using $<\ >$ in figure).

In order to compute the next departure and trace times, we take the minimum of the Lagrangian agreement duration, and the forward and trace CTPs (after subtracting their respective departure and trace times). This minimum, referred to as γ_{map}, is added to current departure-time of forward search and trace-time of trace search (see Fig. 6.7) to get the next times for pair of searches. Correctness of this case is presented in Lemma 6.4 (Sect. 6.3).

Case (b): Forward Search Closed the Destination Node In this case, the algorithm ignores the data collected by the trace search and sets the next forward search departure-time to the earliest of observed intersection points (same as done by CTAS). The next trace-time of the trace search is set to one more than the current shortest path arrival time found by the forward search. Correctness of this case is formally stated in Lemma 6.5 (Sect. 6.3).

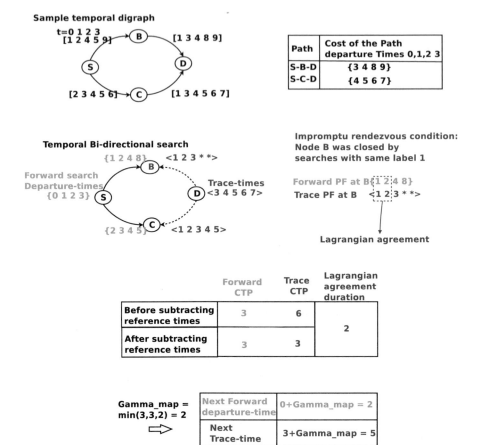

Fig. 6.7 Illustration of the lagrangian agreement duration, γ_{map} in case of impromptu rendezvous

An detailed pseudocode of the BD-CTAS algorithm is given in Algorithm 7. We now provide an execution trace of the algorithm on the ALSP instance shown in Fig. 6.1.

Execution Trace of BD-CTAS Algorithm An execution trace of the BD-CTAS algorithm is shown in Fig. 6.8. With the intention of keeping the figure clear, the inner details of the priority queues are not shown. As mentioned earlier, the priority queue of forward search is similar to that of CTAS. While, the priority queue in the trace search is ordered on the absolute value of the difference between the trace-time at the destination and the trace function values. Figure 6.1 only gives a high-level conceptual trace of BD-CTAS algorithm for the first iteration for ALSP instance. Here, a pair of forward (from node S with departure-time $t = 0$) and trace searches start (from D with trace-time $t = 7$) compute a single sub-interval optimal Lagrangian path.

Algorithm 7 Bi-directional CTAS algorithm

Input: a temporal digraph *STN*, source *S*, destination *D*, a departure-time interval $\lambda = [t_{d_1} \ldots t_{d_k}]$

1: Build a reverse temporal digraph STN_{rev} with each edge in *STN* reversed.
2: Set t_{a_1} and t_{a_m} to arrival time along the fastest path between *S* and *D* for times t_{d_1}, and t_{d_k}
3: Set the trace-time interval for the trace search $\lambda_{rev} = [t_{a_1} \ldots t_{a_m}]$
4: Set the current departure-time of forward search $for_{st} = t_{d_1}$ and of trace search $tra_{st} = t_{a_1}$
5: **while** a shortest path for each time $t \in \lambda$ waits to be determined **do**
6:     Initialize priority queues PQ_{for} with path functions corresponding to *S* and its neighbors in *STN*
7:     Initialize priority queue PQ_{tra} with trace functions corresponding to *D* and its neighbors in STN_{rev}
8:     $FP^{min} \leftarrow$ ExtractMin from PQ_{for}
9:     $TP^{min} \leftarrow$ ExtractMin from PQ_{tra}
10:     Add the tail node of FP^{min} and TP^{min} to set of closed nodes
11:     Test the *impromptu* rendezvous condition
12:     **while** *D* is not closed by forward search *or* impromptu rendezvous condition is not satisfied **do**
13:         Determine the earliest intersection point of FP^{min} with other path functions in PQ_{for}
14:         Save the intersection point in forward path intersection table
15:         Expand FP^{min} and push the newly determined path functions into PQ_{for}
16:         Delete paths ending on tail of FP^{min} from PQ_{for}
17:         $FP^{min} \leftarrow$ ExtractMin from PQ_{for}
18:         Determine the earliest intersection point of TP^{min} with other trace functions in PQ_{tra}
19:         Save the intersection point in trace path intersection table
20:         Expand TP^{min} and push the newly determined trace functions into PQ_{tra}
21:         Delete all the paths ending on tail node of TP^{min} from the PQ_{tra}
22:         $TP^{min} \leftarrow$ ExtractMin from TQ_{for}
23:         Add the tail node of FP^{min} and TP^{min} to set of closed nodes
24:         Test the impromptu rendezvous condition
25:     **end while**
26:     Compute γ_{map}
27:     **if** impromptu rendezvous condition was satisfied **then**
28:         $for_{st} = for_{st} + \gamma_{map} + 1$ /*Next departure-time for the forward search */
29:         $tra_{st} = tra_{st} + \gamma_{map} + 1$ /*Next time instant where the trace search starts */
30:     **else**
31:         $for_{st} \leftarrow$ earliest intersection point in the forward path intersection table.
32:         $tra_{st} = 1 +$ arrival time of fastest path found for time $t = (for_{st} - 1)$
33:     **end if**
34: **end while**

In the first step (after closing *S* and *D*) the forward and trace searches add the path functions to their corresponding neighbors to their respective priority queues. Path functions are indicated using {} or (< >) next to the node. For instance, the forward path function of *S-A* is {1 2 8 9} and the trace path function of *D-F* is < 4 5 6 7 8 >.

In the second step, the forward search closes node *A*, and the trace search closes node *F*. The intersection points (among the path functions) encountered at this stage are also shown in Fig. 6.8. Figure 6.9 shows the optimal paths and their corresponding path functions for all the closed nodes in every step. Now, paths *S-A* and *D-F* are expanded to include the neighbors.

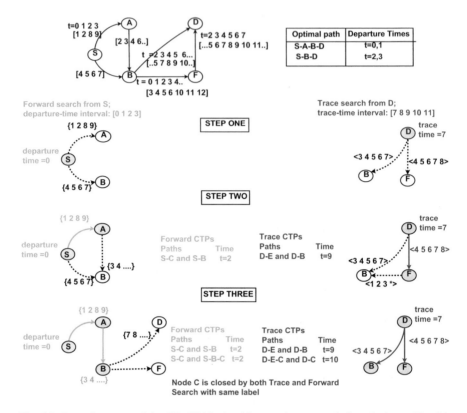

Fig. 6.8 Execution trace of the BD-CTAS algorithm on the temporal digraph shown Fig. 6.1a (replicated on *top left*). Forward search is shown in *green* and trace search is shown in *blue*. Nodes which are closed by either search have *shaded* background

Node	STEP ONE Forward Search Closes	STEP ONE Trace Search Closes	STEP TWO Forward Search Closes	STEP TWO Trace Search Closes	STEP THREE Forward Search Closes	STEP THREE Trace Search Closes
S	Path Found: S Label 0 PFunc{0 1 2 3}					
D		Path Found: D Label 7 PFunc[7 8 9 10 11]				
A			Path Found: S-A Label 1 PFunc{1 2 8 9}			
F				Path Found: D-F Label 4 PFunc<4 5 6 7 8>		
B					Path Found: S-A-B Label 3 PFunc {3 4 ...}	Path Found: D-B Label 3 PFunc <3 4 5 6>

Fig. 6.9 Corresponding optimal paths and paths functions for closed nodes in Fig. 6.8

In the following step both forward and trace searches close node B with the same label 3 (see Fig. 6.9). This happens to be our impromptu termination condition. Now, the algorithm would juxtapose the forward path function ({3 4 …}) and the trace function < 3 4 5 6 7 >) of node B to determine the Lagrangian agreement duration. After combining with the forecasted forward and trace CTP information, the next departure-time of the forward search would be set to $t = 2$, while the next trace-time of the trace search would be set to $t = 7 + 2 = 9$. With this the algorithm has successfully computed one sub-interval optimal Lagrangian path with and $\omega = 0, 1$. This process continues until fastest paths for all departure-times are found.

6.3 Analytical Analysis of Temporal Bi-Directional Search

In this section, we state the formal claims which are need to prove the correctness and completeness of the BD-CTAS algorithm. These claims are reproduced from [29] for sake of convenience of the reader. Interested readers are encouraged to read [29] for full details.

6.3.1 Correctness of BD-CTAS Algorithm

Correctness claim of BD-CTAS algorithm has four key components. In the first part, we need to prove that the forward search is correct and complete. This was established in Sect. 5.4 of Chap. 5. Following this, we need to establish that the trace search is correct and complete. Mathematical implication of this statement is given Lemma 6.1. In the third part, correctness of the impromptu rendezvous termination condition is established. This implies that when BD-CTAS algorithm terminates on the impromptu rendezvous condition, we have a path which is optimal for at least one departure time. This is established in Lemma 6.3 (using Lemma 6.2). And finally, given the optimality of path for one departure time, the algorithm then forecasts the next departure time for the forward and trace-time for the trace search. Lemmas 6.4 and 6.5 are used to prove the correctness of this part. Using all of these four parts, Theorem 6.1 establishes the correctness of BD-CTAS for any instance of the ALSP problem.

Lemma 6.1 *Assume an instance of the trace search which started from node d (which is the destination) at time t_{arr} over the reverse of the input temporal digraphs. Further assume a stage where a node u is closed by this search and t_α is the earliest of the trace critical time point observed so far. If P: d-v_1-…-v_k-u is the path determined by this search (to node u) and Q is the path with edges in P reversed; then we have the following*

- Q is the fastest path between u to d for all arrival times in $t \in [t_{arr} \ (t_\alpha + t_{arr})]$ at d. By fastest we mean, Q has latest departure among all paths for the same arrival time $t \in [t_{arr} \ (t_\alpha + t_{arr})]$.

Proof Detailed proof is given in [29].

Lemma 6.2 *Assume an instance of the bi-directional search on a temporal digraph with forward search exploring from the source at time t_{dep}, and trace search exploring from the destination at time t_{arr}. Both the searches can close a node u with the same distance label iff $t_{arr} \geq a_{optimal}$. Here, $a_{optimal}$ is the time at which the forward search would close the destination node.*

Proof Detailed proof given in [29].

Lemma 6.3 *Assume an instance of the bi-directional search on a temporal digraph with forward search exploring from the source at time t_{dep}, and trace search exploring from the destination at time t_{arr}. It is sufficient to terminate the searches as soon as both of them close a node v with the same distance label, only iff $t_{arr} \leq a_{optimal}$. Here, $a_{optimal}$ is the time at which the forward search would close the destination node.*

Proof Detailed proof given in [29].

Lemma 6.4 *Assume an instance of the bi-directional search on a temporal digraph with forward search exploring from the source at time t_{dep}, and trace search exploring from the destination at time t_{arr}. Here, t_{arr} is guaranteed to be $\leq a_{optimal}$. If the searches terminate using the* impromptu rendezvous *condition with a $\gamma_{map} = x \ (x \geq 1)$, the we have the following:*

1. *The algorithm correctly determines the shortest path for departure-times t_{dep} through $t_{dep} + x - 1$;*
2. *A trace-time of $t_{arr} + x$ is not an overestimate of optimal arrival time for a forward search starting at time $t = t_{dep} + x$.*

Proof Detailed proof given in [29].

Lemma 6.5 *Assume an instance of the bi-directional search on a temporal digraph with forward search exploring from the source at time t_{dep}, and trace search exploring from the destination at time t_{arr}. If the algorithm terminates with destination node being closed and $t_{d_i} + x$ being the forecasted forward CTP, then we have the following:*

1. *The algorithm correctly determines the fastest path for departure-times t_{dep} through $t_{dep} + x - 1$;*
2. *The value $PFun_P(t_{dep} + x - 1) + 1$ is not an overestimate of the arrival time for a forward search starting at time $t = t_{dep} + x$. Here, $PFun_P(t_{d_i} + x - 1)$ is the arrival time at the destination along the current fastest path P found by the forward search for the departure time $t = t_{dep} + x - 1$*

Proof Detailed proof given in [29].

Theorem 6.1 *Given an instance of the ALSP problem with a source S, a destination D and, a discrete departure-time interval (at S) $\lambda = [t_1 \ t_2 \ t_3 \dots t_k]$, the BD-CTAS algorithm correctly computes the solution for all the departure-times given in λ.*

Proof Detailed proof given in [29].

6.4   Conclusion

Bi-directional search on temporal digraphs is challenging as we do not know the time-instant at which the exploration should start from the destination node. Current state of the art addresses this challenge by executing the backward search on the lower bound graph. However, we can elevate this challenge in the ALSP problem by computing the optimal arrival times for the first and the last departure-times in the given departure-time interval λ. We can use these times in the backward search. In the resulting algorithm, the backward search (now known as *trace search*) would now be running on the reverse of the input temporal digraph itself.

Chapter 7
Knowledge Discovery: Temporal Disaggregation in Social Interaction Data

Ritvik Shrivastava and Sreyashi Nag

Given the ever increasing penetration of communication technologies, it is increasingly becoming feasible to observe and study social interaction data. For instance, it is now possible to answer questions such as "Which communities are transient in nature, i.e., they exist for a short period of time?", "How does an individual members social capital (as measured through his/her centrality) fluctuates over time?", etc.

Much of the traditional work [1, 67] in the area of social networks assumed that the underlying social network is does not change with time. However, this is not true in many cases. For example, consider the evolution of an individual's friendship network in an online social networking platforms such as Facebook (www.facebook. com) as he/she moves through different phases professional life (undergraduate, graduate, and full-time job). It is conceivable that strength of some of the ties may go down with time. Similarly, from time to time, we may form some short-lived groups geared towards a specific task, for e.g., a group of faculty getting together for a period of few months over a project proposal. During this time, they may be interacting with each other more frequently than with others. One can find several such real world scenarios, where social aspects of our lives would change with time. As a consequence, it raises the need for developing computational techniques which can consider the time-varying nature of social structures.

To this end, several researchers have started working on this aspect of social interaction data. These works can be broadly classified into following four categories:

- Shortest path centrality metrics for dynamic social networks [28, 30, 40, 41, 61]: These works have extended the traditional shortest path based centrality metrics such as Betweenness [3, 18, 19] and Closeness [19] for temporal digraphs.
- Random walk based centrality for dynamic social networks [47]: This work generalized α-centrality for dynamic networks. The key idea of this work was to adapt the random walk such that it gives more weight to interactions which are closer in time. This way it avoids exploring unnecessarily long paths created by

© Springer International Publishing AG 2017

V.M.V. Gunturi, S. Shekhar, *Spatio-Temporal Graph Data Analytics*,

https://doi.org/10.1007/978-3-319-67771-2_7

stitching interactions over very long periods of time, for example, an interaction between individuals X and Y in Jan 2017 would most likely not be related to an interaction between Y and Z in Jan 2018.

- Persistent community detection in temporal digraphs [6, 62–64]: These works were one of the first to start considering community detection in a time-varying scenario. The central premise of these works being, each individual would have his "primary" group where he would spend most of his time (i.e., most of his interactions). And other "outside" interactions would not amount to much. Along those lines, the algorithms developed in [6, 62–64] focused on determining these "primary" groups.
- Transient community detection in temporal networks [33, 46]: Both these works perform a content-based aggregation of individuals into communities. In other words, they can effectively detect communities surrounding a particular news article or topic.

This chapter presents a temporal generalization of a random-walk based technique for community detection on a temporal digraph representation of the social interaction data. This algorithm can discover social structures at both lower temporal granularities (e.g., daily, weekly) as well as higher levels of temporal granularity (e.g., yearly). The key idea in the adaptation is that as a random walk progresses over a temporal digraph (*temporal random walk*), we would first come across transient clusters which would then disperse as the random walk starts considering the network at larger time granularities. In other words, as the random walk progresses over time, only the permanent clusters, i.e., clusters that persist for a significant time period (and in turn, long term centrality metrics for individual nodes), are likely to remain. This technique is capable of detecting transient as well as persistent communities in social interaction data. In addition to community detection, this chapter also presents a random walk based technique for computing a temporal adaptation of the katz centrality [38].

Outline of Rest the Chapter Section 7.1 presents the basic concepts related to temporally detailed social networks (TDSN), transient and persistent communities. In Sect. 7.2, we discuss the formation of temporal paths in TDSNs. We present a random walk based technique to detect transient communities in Sect. 7.3. Temporal adaptation of Katz centrality is discussed in Sect. 7.4. In Sect. 7.5, we present a case study highlighting the difference between random walk based techniques developed for TDSNs and their traditional counterparts in static graphs.

7.1   Basic Concepts and Problem Definition

A Temporally Detailed Social Network (TDSN) is a collection of panels, where each panel is in turn an aggregation of all the social interactions which happened over a window of fixed length, e.g., minute, hour, day, week, month, etc. Length of this window, hereafter referred to as *TDSN-granularity*, and is decided according

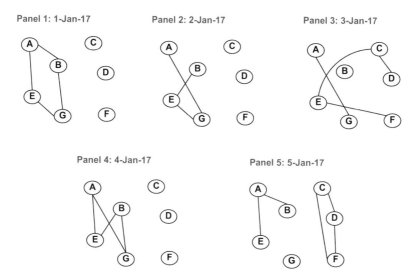

Fig. 7.1 Sample social interaction data

to the needs of the application domain. Figure 7.1 illustrates a sample TDSN consisting of five panels. Note that if two individuals X and Y interact with each other more than once inside a window, then TDSN would still consist of only one edge between nodes X and Y, however, its weight would be different. Note that it is important to keep the value of *TDSN-granularity* as less as possible. A higher value of this parameter would wrong estimation of number of paths where edges are temporally ordered. Implicitly, we are assuming that all the interactions inside one panel happened at the "same instant".

Weight of an Edge in a Panel Inside a panel which is an aggregation of all the social interactions that happened over n time units, we compute the weight of an edge P_{ij} between two nodes i and j as follows:

$$P_{ij} = \frac{\text{\# times `i' was connected to `j'}}{\text{Length of the panel}} \quad (7.1)$$

Communities Traditionally, a community has been loosely defined as a collection of individuals who happen to be similar to each other in some sense. This similarity could arise from having same set of interests, being a part of the same organization, etc. Also, it is generally expected that the individuals would have stronger levels of interaction with people from the same community than with individuals outside the community. In case of temporally detailed social networks, the notion of communities can be further divided into: (a) transient communities and, (b) persistent communities.

Transient Community in a TDSN (C_t) is defined as a group of individuals who interact with each other more often during a short time period. Inside this time interval, these group of individuals interact with each other more often than with others. This communication ceases to exist at other times during in the time horizon of the TDSN. Thus, transient communities have a temporal locality. The members of the transient community need not be in each other's neighborhood and consequently, need not have had any interaction in the past.

Persistent Community in a TDSN (C_p) is defined as a group of individuals who in general interact with each other more than they interact with others. And this level of interaction is maintained throughout the time horizon of the TDSN. Persistent communities have strong interactions between its members by virtue of some level of similarity between them. This similarity is responsible for its prolonged duration during the life span of the network. Transient communities, on the other hand, are possibly created to achieve a certain goal. For example, a group of researchers getting together to write a project proposal. In a TDSN, random-walk based techniques are based on the concept of temporal paths. The observations for these techniques can be made at variable time granularities. This concept is discussed below.

Temporal Paths A temporal path P in a TDSN is a sequence of nodes $P = (v_1, v_2, v_3 \ldots v_n, v_n + 1)$, where each edge in P belongs to a different panel in the TDSN. Moreover, the panel number of (v_i, v_{i+1}) is strictly less than the panel number of (v_{i+1}, v_{i+2}). In other words, a temporal path can be defined as a sequence of temporal edges.

Problem Definition
Input: A collection of panels representing a temporally detailed social network (TDSN).
Output: An algorithm framework for evaluating the presence of transient communities as well as assessing the influence measures of individual nodes.
Objective: Extracting information which can otherwise not be evaluated using traditional methods.

7.2 Formation of Temporal Paths in TDSN

Consider an adjacency matrix M representation of any connected graph, each cell denotes the presence or absence of a direct edge (connection) between each node pair. The cell would contain 1 if there is an direct edge between the pair of nodes, otherwise it would be 0. For such a matrix M, the matrix M^n would contain the number of *n-hop* paths between each pair of nodes. In other words, after multiplying the matrix M n times, each cell (i, j) would contain the number of paths between

Fig. 7.2 0-hop (i.e., direct connection) and 1-hop path formation

M	P	Q	R	S
P	0	1	0	0
Q	0	0	1	0
R	0	1	0	1
S	1	0	0	0

MXM	P	Q	R	S
P	0	0	1	0
Q	0	1	0	1
R	1	0	1	0
S	0	1	0	0

M_1	A	B	C	D
A	0	1	0	0
B	0	0	1	1
C	0	1	0	1
D	1	0	0	0

M_2	A	B	C	D
A	0	1	1	0
B	1	0	0	1
C	0	1	0	1
D	1	0	0	0

M_1XM_2	A	B	C	D
A	1	0	0	1
B	1	1	0	1
C	2	0	0	1
D	0	1	1	0

Fig. 7.3 Constructing temporal paths (of length less than or equal to 1) formation in a TDSN

M_1	A	B	C	D	E
A	0	0.9	0.2	0	0
B	0	0	0	0	0
C	0	0	0	0.9	0
D	0	0	0	0	0
E	0	0	0	0	0

M_1	A	B	C	D	E
A	0	0	0	0	0
B	0	0	0	0.9	0
C	0	0	0	0	0.2
D	0	0	0	0	0.9
E	0	0	0	0	0

M_1	A	B	C	D	E
A	0	0	0	0.81	0.04
B	0	0	0	0	0
C	0	0	0	0	0.81
D	0	0	0	0	0
E	0	0	0	0	0

Fig. 7.4 Constructing weighted temporal paths (of length less than or equal to 1) in a TDSN

nodes i and j which have n nodes or less. This is illustrated in Fig. 7.2 where MxM is a single multiplication which would contain the total number of *0-hop and 1-hop* paths in the network.

We now discuss the formation of temporal paths in temporally detailed social networks (TDSN). For sake of simplicity and easy interpretation, in the following discussion, we assume that the weight of every edge in the panels is 1.

Given a collection of panels as a TDSN, we can obtain temporal paths by the multiplication of two or more panels. If $M_1, M_2, \ldots, M_n$ represents the n panels of a TDSN, the resultant matrix $M_t = M_1 \times M_2 \times \ldots \times M_n$ represents the number of temporal paths (refer Sect. 7.1 for formal definition) of length less than or equal to n, i.e. all temporal paths with at most n intermediate hops between the node pairs. Figure 7.3 illustrates this concept. In this figure, M_1 and M_2 are two consecutive panels in a TDSN. Consider, nodes A and D in the figure. There is no direct connection between them at both snapshots. However, there exists a *1-hop* temporal path A→B→D between them. Note that this process can be easily modified to take into consideration the concept of waiting-time at nodes separately.

Considering the Weight of Edges in Panels Recall the definition of weight of an edge in Sect. 7.1. One can naturally extend the previously discussed notion of multiplication of panels to consider the weights of the edges. The temporal paths now created would have an implicit notion of strength of connection between two nodes. For instance, consider the panels (belonging to a TDSN) shown in Fig. 7.4. In panel M_1 of this network, A was very active in sending emails to B, but not so much with node C. And in panel M_2 of this network, B actively sent emails to D, where C was not so much active in writing emails to E. Thus, when we construct $M_1 \times M_2$ while consider the edge weight, cell corresponding to (A, D) would have a much higher value, implying that the path $A \rightarrow B \rightarrow D$ was stronger than $A \rightarrow C \rightarrow E$. Note that, in this example, matrices corresponding to panels were not symmetric. Such aspects depend on the nature of the data available. In case of emails, the matrices would not be symmetric. Whereas, if the dataset contained phone calls, then the matrices corresponding to the panels would have to be symmetric.

7.3   Community Identification in TDSNs

The key idea over here is to start a random-walk at any node in the TDSN and continue on to the nodes in the next panel. For instance, if nodes x and y are connected in panel p_i, then the temporal random walk would connect this edges with a out going edge from y, say to node z, in panel p_{i+1}. This makes a temporal random walk $x \rightarrow y \rightarrow z$ across the panels p_i and p_{i+1}. Such random walks are likely to remain within the same cluster as the paths increase in length.

Given a series of panels of a temporally detailed social network, communities are detected in the following two steps:

Temporal Spreading   In this step, consecutive panels of the TDSN are multiplied sequentially to produce temporal paths. The step produces longer temporal paths with each multiplication. Nodes that are a part of longer paths are likely to be a part of the same community. Since the algorithm uses the weights of the edges while multiplying (instead of just using presence or absence of an edge), this also ensures that only communities with a decent frequency of interaction are discovered. If M_n denotes collection of edges (represented as a matrix) in panel n, w_1 denotes the lower panel number of the TDSN, and w denotes the number of panels, then result of temporal spreading can be expressed as:

$$TempSpread = \prod_{n=w_1}^{w_1+w} M_n \qquad (7.2)$$

Amplification   In this step, the matrix obtained after *temporal spreading* is 'amplified' to strengthen the strong connections and weaken the feeble ones. Here, the value in each cell in the resultant matrix is raised to the αth power (amplification factor). In other words, this step brings the stronger connections into prominence to be identified as clusters/communities and dilutes the weakly connected edges in the network. Thus implicitly, after this step, the algorithms discards noise in the network that may appear as communities themselves.

Cell (i,j) in Amplified Matrix $= \{value\ of\ Cell\ (i,j)\ in\ TempSpread\}^{\alpha}$ $\qquad (7.3)$

Following these two steps, communities can be detected in the resultant panel by observing non-zero values in each node row. This procedure is described in the following section.

7.3.1 Approach for the Detection of Transient Communities

The community detection approach described next uses a sliding window (of size ω) to determine communities are different temporal resolutions. The size of the sliding window could vary according to the needs of the social question being investigated. For instance, if the user is interested in uncovering the communities at the temporal resolution of one month; then the length of this sliding window would be number of panels corresponding to 1 month. Similarly, if the user is interested communities which lasted for about 6 months, then the length of this window would be number of panels corresponding to 6 months. The steps of the approach are detailed next:

1. *Step 1: Temporal Spreading*: Given the panels of the TDSN and the sliding window size ω as input, temporal spreading is carried out for all the panels which fall in the window as it slides over all the panels of the input temporally detailed social network.
2. *Step 2: Amplification*: After each instance of temporal spreading, amplification is carried out according to the given amplification factor α.
3. *Step 3: Cluster Extraction*: Following each amplification, all the likely clusters that have been detected at that stage are extracted out and stored as *transient communities*. This is done by extracting for every node row, the nodes with non-zero values. One may also choose to have to some thresholds while choosing the nodes. A very low value in the row may not be an indicator of strong cluster.

After Step 3, one can increase the sliding window size ω and repeat steps 1–3 to get transient communities at larger temporal resolutions.
Optional Step 4: Communities at Persistent Level: This can determined when ω is set to the total number of panels in the given TDSN. At this stage, the communities that are extracted using steps 1–3 can be stored as *persistent communities*. Algorithm 8 presents the steps involved in the detection of the set of transient and persistent communities present in the input TDSN.

7.3.2 Running Example for Transient Community Detection

Figure 7.5 illustrates the discussed algorithm on a sample synthetic dataset. This dataset consists of a 7-node TDSN over a time period of 200 timestamps. The snapshots were aggregated into 20 panels, each consisting of ten snapshots. Weights of the edges inside the panel were computed using the equation discussed in Sect. 7.1. Steps 1–4 (optional) are shown in figure.

Algorithm 8 Community detection in temporally detailed social networks

---

1: $\alpha \leftarrow$ Amplification Factor
2: $w_{lower} = 1$ /* w_{lower} is set to the first panel in TDSN */
3: **while** $\omega \leq$ number of panels in TDSN **do**
4:     **while** $w_{lower} + \omega \leq$ last panel of TDSN **do**
5:         $TempSpread = TemporalSpread(w_{lower}, \omega)$ /*Multiplies panels w_{lower} through $w_{lower+\omega}$ */
6:         $Amp = Amplify(TempSpread, \alpha)$
7:         Extract communities from Amp
8:         $w_{lower} = w_{lower} + 1$ /* Window is slid by 1 panel */
9:     **end while**
10:     $\omega = \omega + 1$ /* Increase the resolution of interest by increasing the length of the sliding window */
11: **end while**
12: Return all transient and persistent communities

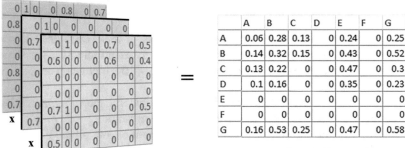

$=$

	A	B	C	D	E	F	G
A	0.06	0.28	0.13	0	0.24	0	0.25
B	0.14	0.32	0.15	0	0.43	0	0.52
C	0.13	0.22	0	0	0.47	0	0.3
D	0.1	0.16	0	0	0.35	0	0.23
E	0	0	0	0	0	0	0
F	0	0	0	0	0	0	0
G	0.16	0.53	0.25	0	0.47	0	0.58

Step 1: Temporal
Spreading with factor $= 3$

	A	B	C	D	E	F
A	0	0.08	0.02	0	0.06	0
B	0	0.1	0.02	0	0.18	0
C	0	0.05	0	0	0.22	0
D	0	0.03	0	0	0.12	0
E	0	0	0	0	0	0
F	0	0	0	0	0	0

Step 2: Result of Amplification with
factor $= 2$

	A	B	C	D	E	F	G
A	0.69	1.29	0	0	1.14	0	1.18
B	0.69	1.29	0	0	1.14	0	1.18
C	0	0	0	0	0	0	0
D	0	0	0	0	0	0	0
E	1.21	2.26	0	0	2	0	2.06
F	0	0	0	0	0	0	0
G	1.47	2.75	0	0	2.43	0	2.5

Step 4: Result at Persistent Level

	A	B	C	D	E	F	G
B	0	0.1	0.02	0	0.18	0	0.27

Step 3: Cluster Extraction corresponding to node row B.
Nodes that are part of the extracted cluster have been highlighted.

Fig. 7.5 Running example of community detection in a TDSN

7.4   Node Influence Measures in TDSNs

In this section, we generalize Katz Centrality for temporally detailed social networks. This would help in studying the "influence" of nodes in the TDSN.

7.4.1   Temporal Katz Centrality

Traditional Katz Centrality measures the total number of direct and indirect paths of all lengths incident on a particular node [38]. This is done while attenuating the effect of longer paths. Following is a mathematical definition of the Katz centrality on static graphs.

$$x_i = \sum_{k=1}^{\infty} \sum_{j=1}^{N} \gamma^k (A^k)_{ji} \tag{7.4}$$

Here, γ is the attenuation factor between 0 and 1, and N is the number of nodes in the input graph. This centrality measure indicates the relative influence of a node in the network.

We now present an temporal adaption of this metric for TDSNs. The adaptation operates under the assumption that as the length of the path increases, the impact of the information traveling through it is reduced. This is taken into account when observing the number of viable temporal paths ending at the node under consideration. In order to account for this, the algorithm attenuates the contribution of longer temporal paths, similar to what the traditional Katz Centrality does in static graphs.

Algorithm 9 presents a pseudocode to compute a temporal adaptation of the Katz centrality on TDSNs. The algorithm takes the following three inputs: (a) initial panel t_i, (b) last panel to consider t_f, (c) window size ω. By choosing setting appropriate values to these parameters, the algorithm can answer a wide variety of these questions. Following are few examples:

1. *Case 1:* In how many different ways can information reach a node through temporal paths of length 1 day between May 1 and August 1? For answering this question: $t_i = 1^{st}$ panel corresponding to May 1, t_f = last panel corresponding to August 1, $\omega = 1$ *day* and the attenuation factor γ should be set to 1.
2. *Case 2:* In how many different ways can information reach a node through temporal paths of length 1 day for the entire duration of the network's existence? For answering this question: $t_i = 1^{st}$ panel in TDSN, t_f = last panel in TDSN, $\omega = 1$ *day* and the attenuation factor γ should be set to 1.
3. *Case 3:* In how many different ways can information reach a node through temporal paths of all lengths between January 1 and February 1? For answering

---

Algorithm 9 Katz centrality for temporally detailed social networks

---

1: $t_i \leftarrow$ panel in TDSN from where analysis starts /*Given as input */
2: $t_f \leftarrow$ panel in TDSN from where analysis stops /*Given as input */
3: $\gamma \leftarrow$ Attenuation Factor /*Given as input */
4: $\omega = 1$ /*Length of the sliding window, incremented after each iteration */
5: $w_{lower} = t_i$
6: **while** $\omega \leq t_f - t_i$ **do**
7:    **while** $w_{lower} + \omega \leq t_f$ **do**
8:       $M = TemporalSpread(w_{lower}, \omega)$ /*Multiplies panels w_{lower} through $w_{lower+\omega}$*/
9:       $K_i = \gamma^\omega \sum_{j=1}^{n} M_{i,j}$ /* n is the #distinct nodes in TDSN*/
10:       $K = K + K_i$
11:       $w_{lower} = w_{lower} + 1$
12:    **end while**
13:    $\omega = \omega + 1$
14: **end while**

---

this question: $t_i = 1^{st}$ panel corresponding to Jan 1, $t_f =$ last panel corresponding to Feb 1, $\gamma = 1$ $\omega = 1 - day, 2 - days, 3\ days, \ldots$, time horizon of the input TDSN.

4. *Case 4:* In how many different ways can information reach a node through temporal paths of all lengths for the entire duration of the network's existence? For answering this question: $t_i = 1$st panel in TDSN, $t_f =$ last panel in TDSN, $\gamma = 1$ $\omega = 1 - day, 2 - days, 3\ days, \ldots$, time horizon of the input TDSN.

Algorithm 9 presents a procedure for calculating temporal katz centrality for cases 3 and 4 only. Cases 1 and 2 can be computed in a straightforward fashion by setting the value of ω appropriately instead of incrementing it in each iteration as done by Algorithm 9.

7.5 Case Study on TDSN

In this section, we present a case study which illustrates the key nature of temporally detailed social networks. More specifically, this case study highlights the fact that social structures (community and social capital) observed in smaller temporal resolutions (e.g., month) are different from the ones seen in larger resolutions (e.g., years). Following two datasets are used in the case study:

1. **University Email Dataset:** This dataset contains email communications from a large European university. This dataset recorded about 161,227 email communications among 4554 individuals. It has been used for multiple studies previously [28, 49, 50]. These email communications were recorded in the following format: $< A, B, t >$, where A and B are two individuals in the University and t denotes the time-stamp (in minutes) at which individual A sent an email to individual B.

2. **CollegeMsg Dataset [58]:** This dataset comprises of private messages sent on an online social network at the University of California, Irvine over a duration of 193 days. Users could search the network for others and then initiate conversation based on profile information, hence creating an edge $< u, v, t >$, where user u sent a private message to user v at time t.

7.5.1 Metric used for Experimentation

Jaccard Index The Jaccard index, also known as the Jaccard similarity coefficient, is a statistic used for comparing the similarity of sets. The Jaccard coefficient measures similarity between finite sample sets, and is defined as the cardinality of the intersection divided by the size of the union of the sample sets:

$$J(A, B) = \frac{|A \cap B|}{|A \cup B|} = \frac{|A \cap B|}{|A| + |B| - |A \cap B|} \tag{7.5}$$

A lower value of the Jaccard index implies that the sets A and B (or communities in this case) are dissimilar from each other. In our case, a lower Jaccard score while comparing persistent communities to communities at smaller temporal resolution signifies the presence of a transient groups. In this case study, the Jaccard Index is obtained for each node in the network by comparing every transient group discovered at each time granularity with the persistent community obtained at the persistent level. For example, consider a network of 6 nodes A, B, C, D, E, F. Assume that at the level of persistent communities nodes A, B and E form a community. Whereas, at some intermediate temporal resolutions (i.e., at lower values of ω), assume that nodes A, C and F form a community as per our algorithm. If the set of nodes corresponding to the persistent community of node A is set P_A, and the transient ones is set T_A. We calculate Jaccard(P_A, T_A) as $\frac{sizeOf(P_A \cap T_A)}{sizeOf(P_A \cup T_A)}$. The set $P_A \cup T_A$ consists of nodes A,B,C,E,F, whereas the set $P_A \cap T_A$ consists of only node A. The resultant Jaccard(P_A, T_A) value for node A would be $1/5 = 0.2$. This is represented through a confusion matrix in Fig. 7.6.

Both in Transient and Persistent	In Transient but Not in Persistent		1	2
Not in Transient but in Persistent	In Neither (Irrelevant Nodes)		2	Irrelevant

$$\text{Jaccard}(P_A, T_A) = \frac{sizeOf(P_A \cap T_A)}{sizeOf(P_A \cup T_A)} = \frac{1}{1+2+2} = \frac{1}{5} = 0.2$$

Fig. 7.6 Confusion matrix representation of Jaccard index

7.5.2   Experiment 1

Hypothesis Community structures present at smaller temporal resolutions are different than the ones present at the final level—where ω is set to entire time horizon of the input TDSN.

Experimentation In order to investigate the above hypothesis, we compared the communities detected for different values of temporal resolutions (i.e., length of the sliding window ω) against the ones detected at the persistent level, i.e. the communities detected while viewing the entire TDSN (ω = entire time horizon of the input TDSN). We used the previously discussed Jaccard index to compare these. For a particular value of ω and a particular individual, as the sliding window moves forward, we add the resultant Jaccard scores across all positions of the sliding window in the TDSN to obtain a sum. If T_i is a transient community for node A at one position of the sliding window and P be its persistent community found at top level; then the sum of Jaccard indices for node A across θ positions of the sliding window (for same ω), is calculated using:

$$Sum_i = \sum_{i=1}^{\theta} J(T_i, P) \tag{7.6}$$

A smaller sum indicates a larger difference between the communities detected at lower time granularities when compared to those detected at the persistent level. In other words, a smaller sum value signifies a greater proportion of transient communities detected at that granularity. Figure 7.7 illustrates the results of this experiment. The figure illustrates that, for a large number of individuals (nodes in the figure), there is a huge difference between their transient and their persistent communities.

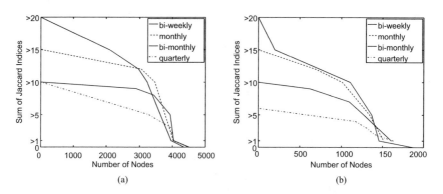

Fig. 7.7 Sum of Jaccard indices calculated for different values of ω for both datasets. (**a**) University Email dataset. (**b**) CollegeMsg dataset

7.5.3 Experiment 2

Hypothesis Information obtained from TDSNs (which implicitly preserve the temporal sequence of interaction events) cannot be obtained by static analysis performed on the aggregated TDSN.

Experimentation To investigate this hypothesis, we compare the communities detected by static analysis on aggregated TDSNs against those detected by our temporal random-walk based approach. For every size of the sliding window (ω), we extracted communities using the approach discussed in this chapter. Following this we created a static graphs by aggregating (collapsing the time dimension) all the panels for every distinct position of the sliding window, as it slides along the panels in the TDSN. On these static graphs, we applied the traditional Markov Clustering Algorithm to extract communities. Using the Jaccard index, we evaluated the difference between the two sets of communities. We calculate the mean Jaccard index for a node across a particular value of ω as a measure of this difference.

Figure 7.8 shows the results of this experiment. One can observe that, in both the CollegeMsg and the University Email datasets, mean Jaccard index for a window length of one quarter is less than that for 2 weeks. This is because of the fact that in case of one quarter, we would be doing more aggregation and thus inflicting more damage to the temporal ordering of interactions. As a result, the communities obtained from aggregated panels would be much more different than the ones seen while respecting the temporal ordering of interactions (as done in the algorithm presented in this chapter). This wide difference shows up as low values of mean Jaccard Index for several nodes in the network.

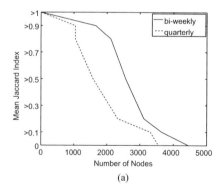
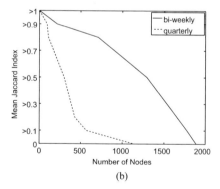

Fig. 7.8 Mean of Jaccard indices for comparing time-aware approach against the MCL algorithm over different temporal resolutions. ω was set to 2 weeks and 1 quarter. (**a**) University Email dataset. (**b**) CollegeMsg dataset

7.5.4 Experiment 3

Hypothesis The influence of nodes measured at distinct temporal resolutions is different from their influence in the aggregated representation of social interactions.

Experimentation In this experiment we compare the result of Temporal Katz centrality presented in this chapter against the traditional Katz Centrality method applied on the aggregated data. For the purpose of presentation, we chose to study Case 3 of the proposed Temporal Katz Centrality as described in Sect. 7.4. The initial time was set to the beginning of the TDSN and the final time as the 2-month ahead of the initial time. The proposed algorithm was run on every window size, between 1 day and 2 months. At each window size the top 1%, 5% and 10% most influential nodes were recorded. This data at every window size was compared against the top 1%, 5% and 10% most influential nodes obtained through the traditional Katz Centrality approach applied on the static graph obtained by aggregating the panels of the window.

Figure 7.9 presents the results of this experiment. As the figure shows, with exception of window length of 1 day, in general there is decrease in agreement as the window length is increased from 2 days to 2 months. This is again intuitive because as we increase the length of window, in case of traditional katz, we would be doing more aggregation and thus inflicting more damage to the temporal ordering of interactions. This results in greater disagreement (thus lower values of Jaccard index) between temporal Katz (computed on TDSN) and traditional Katz (computed on aggregation of panels).

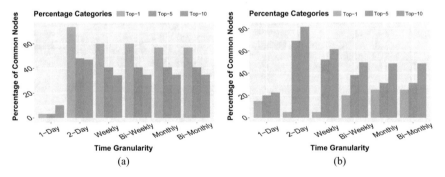

Fig. 7.9 Traditional vs temporal Katz centrality. (**a**) University Email dataset. (**b**) CollegeMsg dataset

7.6 Conclusion

Using the approach presented in this chapter, we observed that the communities seen at smaller time granularities are quite different than the ones seen at larger granularities. We also showed that information obtained by detailed temporal analysis of TDSNs is vastly different from that attained from aggregated versions of the same graph. Lastly, our modified katz centrality calculated a member's influence on the network at smaller time granularities which was seen to differ from the static metric applied at the aggregated scale.

Acknowledgements I would like to thank my students "Ritvik Shrivastava" and "Sreyashi Nag" for their contributions in this work.

Chapter 8
Trend Topics: Engine Data Analytics

Rich instrumentation (e.g, GPS receivers and engine sensors) in modern fleet vehicles allow us to periodically measure vehicle sub-system properties [2, 36, 37] of the vehicle. These datasets, which we refer to as vehicle measurement big data (VMBD), contains a collection of trips on a transportation network. Here, each trip is a time-series of attributes such as vehicle location, fuel consumption, vehicle speed, odometer values, engine speed in revolutions per minute (RPM), engine load, emissions of greenhouse gases (e.g., CO_2 and NOx), etc. Figure 8.1 illustrates a sample VMBD in form of a table. Computationally, engine measurement big data can be represented as a spatio-temporal graph. It should be noted that this is still open area of research. We now present two sample problems which can defined on this data.

8.1 Statistically Significant Hot Route Discovery

Given a spatio temporal graph representation of vehicle measurement big data (VMBD), the goal of statistically significant hot route discovery is to identify simple paths in the underlying spatio-temporal graph which have a significantly higher instances of discrepancies (e.g., deviation from US EPA standards (US Environmental Protection Agency 2015)) in a given target engine variable (e.g, NOx or $CO2$) than the other parts of the spatio-temporal graph.

Finding such specialized hotspots of discrepancies could spur next level of research questions such as: What is the scientific explanation behind a certain high emission observation? Was it due to a particular acceleration-breaking pattern? Can this pattern be coded into the engine control unit for better fuel economy? Figure 8.2 illustrates NOx hotspots (red portions of the trajectory) on a transportation network for conventional diesel engine using measured engine data from buses in Minneapolis-St Paul area.

© Springer International Publishing AG 2017
V.M.V. Gunturi, S. Shekhar, *Spatio-Temporal Graph Data Analytics*,
https://doi.org/10.1007/978-3-319-67771-2_8

Fig. 8.1 Sample vehicle
measurement big data

Vehicle id	Time stamp	Location (lat, long)	Map-matched edge-id	Attributes		
				NO$_x$	Engine RPM	. . .
V1	08:00:00	44.9778° N, 93.2652° W	E1	0.492	860	
V1	08:00:10	44.9778° N, 93.2650° W	E2	0.492	2,980	
V1	08:00:20	44.9777° N, 93.2649° W	E3	0.532	2,730	
V1	08:00:30	44.9775° N, 93.2648° W	E4	0.352	3,870	
V1	08:00:40	44.9774° N, 93.2647° W	E5	0.452	2,321	
. . .						

Fig. 8.2 Sample VMBD
showing high NOx in left
turns

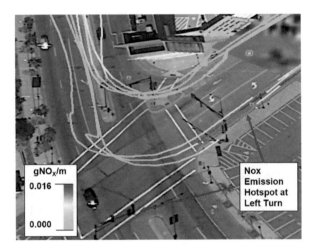

gNO$_x$/m

0.016

0.000

Nox Emission Hotspot at Left Turn

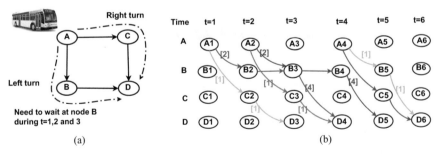

Fig. 8.3 Illustrating hot routes on a Time-expanded graph. (**a**) Two routes between A and D for the bus. (**b**) Each edge is annotated with the number of discrepancies in the target variable (e.g., NOx)

Figure 8.3 illustrates a sample scenario for hot route discovery. Here, a target engine variable (e.g, NOx) is recorded as a bus travels along two different paths between node A and node D in the transportation network containing nodes A, B, C and D (Fig. 8.3a). The path A-B-D involves a left turn at B which takes one unit of time for times t=1,2 and 3 (mimicking rush hours). Whereas, the A-C-D has a right turn which does not involve any waiting. Further assume that the bus takes one time unit to travel on each of the edges in Fig. 8.3a. Figure 8.3b illustrates seven different journeys between A and D on a spatio-temporal graph representation of the transportation network. For sake of simplicity, the figure illustrates only those ST edges which were traveled in some journey. Each journey is shown using a different color. The number mentioned in [] on the edges denote the number of discrepancies in the target variable observed on that edge. As can been seen the journeys in color red and purple (both over the route with left turn during rush hours) have the highest number (six each, compared to others with only two) of discrepancies, making them a potential output for the hot route discovery problem.

Mining statistically significant hot routes is challenging due to the large data volume and the STG data semantics, which violates critical assumptions underlying traditional techniques for identifying statistically significant hotspots [45, 55] that data samples are embedded in isotropic Euclidean spaces and spatial footprints of hotspots are simple geometric shapes (e.g., circles or rectangle with sides parallel to coordinate axes). For our proposed hot route discovery problem, one needs to generalize spatial statistics beyond iso-tropic Euclidean space to model the spatio-temporal graph semantics such as linear routes. Computationally, the size of the set of candidate patterns (all possible directed paths in the given spatio-temporal network) is potentially exponential. In addition, statistical interest measures (e.g., likelihood ration, p-value) may lack monotonicity. In other words, a longer directed path may have a higher likelihood ratio and its sub-paths. We now describe some potential research questions for the problem of hot route discovery.

8.1.1 Potential Next Steps

Step 1: Design Interest Measure for Statistically Significant Hot Routes The goal over here would be to develop interest measures for identifying statistically significant hot routes. This would require balancing conflicting requirements of statistical interpretation and support for computationally efficient algorithms. A statistically interpretable measure is the one whose values would typically conform to a known distribution. On the other hand, computational properties which are known to facilitate efficient algorithm include properties such as monotonicity. For this task one could start by exploring a ratio based interest measure containing the densities of non-compliance inside the path to that of outside the path. We could investigate the merit of this measure through following research questions: What are the characteristics of this interest measure? Does it have monotonicity? If not then can we design an upper bound which has monotonicity? Do the values of this measure (or its upper bound) follow a known distribution? Does it belong to one of the general classes of distributions? If not then how would be an appropriate null hypothesis be designed for monte-carlo simulations? What distribution does the underlying VMBD normally tends to follow (needed for testing null hypothesis)?

Can we derive an interest measure based on the underlying distribution of data itself (similar to log likelihood ratio [45]?

Step 2: Design a Computational Structure of Exploring the Space of Paths   As a first step towards addressing the challenge of large number of candidate paths, one would have to investigate scalable techniques for computing all-pair shortest paths on the spatio-temporal graph representation of the underlying transportation network. This is under the assumption that people (and hence their) traces typically take shortest paths in the network. This would require addressing the challenge of non-stationary ranking among alternative paths between two points in a spatio-temporal graph. This non-stationary ranking precludes the possibility of a dynamic programming (DP) based approach. A potential approach for this challenge could involve a divide-and-conquer strategy called critical-time-point (CTP) based approaches presented in earlier (ref Chap. 5). A CTP based approach addresses the challenge of non-stationary ranking by efficiently dividing the given time interval (over which we observe non-stationary ranking among alternative paths) into a set of disjoint time intervals over which ranking is stationary. One can now use a DP based technique over these sub-intervals. Using the concept of CTPs, one could adapt algorithms (e.g., Floyd Warshalls, Johnsons) for the all-pair shortest path problem.

Step 3: Algorithm for Exploring the Space of all Simple Paths in a Spatio-Temporal Graph   The third step for the hot route discovery would be to investigate the computational challenges of enumerating all the possible paths in spatio-temporal graph (STG). Clearly, in any real spatio-temporal graph, there would exponential number of candidate paths. A naïve approach could involve an incremental spatio-temporal join based strategy, which starts off with singleton edges in the STG and iteratively joins them (based on spatio-temporal neighborhood constraints) to create longer paths. This naïve algorithm raises several unanswered questions: Would be paths enumerated this way would still be simple, i.e., no nodes are repeated? How can we avoid creating loops?

8.2   Conclusion

Given the proliferation of sensors on vehicular engines, it is increasingly becoming possible to get data on various engine related parameters (e.g., speed, torque, power, fuel-input, etc.) as it moves through the city. This data can provide us insights on the behavior of engine under different real life conditions. However, this dataset poses unique challenges to the state of the art in spatio-temporal graph data analytics. New research is needed before we can harness the full potential of this data.

References

1. Aggarwal, C.C. (ed.): Social Network Data Analytics. Springer, New York (2011)
2. Ali, R.Y., Gunturi, V.M., Kotz, A.J., Shekhar, S., Northrop, W.F.: Discovering non-compliant window co-occurrence patterns: a summary of results. In: Advances in Spatial and Temporal Databases. SSTD 2015, pp. 391–410. Springer, Cham (2015)
3. Anthonisse, J.M.: The rush in a directed graph. CWI Technical Report Stichting Mathematisch Centrum. Mathematische Besliskunde-BN 9/71, Stichting Mathematisch Centrum (1971)
4. Batchelor, G.: An Introduction to Fluid Dynamics. Cambridge University Press, Cambridge (1973)
5. Berge, C.: Graphs and Hypergraphs. Elsevier, Oxford (1985)
6. Berger-Wolf, T.Y., Saia, J.: A framework for analysis of dynamic social networks. In: Proceedings of the 12th ACM SIGKDD International Conference on Knowledge Discovery and Data Mining, pp. 523–528. ACM, Philadelphia (2006)
7. Braha, D., Bar-Yam, Y.: From centrality to temporary fame: dynamic centrality in complex networks. Complexity **12**(2), 59–63 (2006)
8. Brandes, U.: A faster algorithm for betweenness centrality. J. Math. Sociol. **25**(2), 163–177 (2001)
9. Chabini, I.: Discrete dynamic shortest path problems in transportation applications: complexity and algorithms with optimal run time. Transp. Res. Rec. J. Transp. Res. Board **1645**(1), 170–175 (1998)
10. Chabini, I., Lan, S.: Adaptations of the A* algorithm for the computation of fastest paths in deterministic discrete-time dynamic networks. IEEE Trans. Intell. Transp. Syst. **3**(1), 60–74 (2002)
11. Dehne, F., Omran, M.T., Sack, J.R.: Shortest paths in time-dependent FIFO networks using edge load forecasts. In: Proceedings of the Second International Workshop on Computational Transportation Science, IWCTS '09, pp. 1–6 (2009)
12. Delling, D., Nannicini, G.: Bidirectional core-based routing in dynamic time-dependent road networks. In: Algorithms and Computation, pp. 812–823. Springer, Berlin/Heidelberg (2008)
13. Demiryurek, U., Banaei-Kashani, F., Shahabi, C.: A case for time-dependent shortest path computation in spatial networks. In: Proceedings of the ACM SIGSPATIAL International Conference on Advances in GIS, GIS '10, pp. 474–477 (2010)
14. Demiryurek, U., Banaei-Kashani, F., Shahabi, C., Ranganathan, A.: Online computation of fastest path in time-dependent spatial networks. In: Advances in Spatial and Temporal Databases. Lecture Notes in Computer Science, vol. 6849, pp. 92–111. Springer, Berlin/Heidelberg (2011)

© Springer International Publishing AG 2017
V.M.V. Gunturi, S. Shekhar, *Spatio-Temporal Graph Data Analytics*,
https://doi.org/10.1007/978-3-319-67771-2

15. Demiryurek, U., Banaei-Kashani, F., Shahabi, C., Ranganathan, A.: Online computation of fastest path in time-dependent spatial networks. In: Proceedings of the 12th International Conference on Advances in Spatial and Temporal Databases, SSTD'11, pp. 92–111. Springer, Berlin/Heidelberg (2011)
16. Ding, B., Yu, J., Qin, L.: Finding time-dependent shortest paths over large graphs. In: Proceedings of the 11th International Conference on Extending Database Technology: Advances in Database Technology, pp. 205–216. ACM, Nantes (2008)
17. Eckmann, J.P., Moses, E., Sergi, D.: Entropy of dialogues creates coherent structures in e-mail traffic. Proc. Natl. Acad. Sci. **101**(40), 143330–14337 (2004)
18. Freeman, L.C.: Centrality in social networks conceptual clarification. Soc. Netw. **1**(3), 215–239 (1979)
19. Freeman, L.C.: A set of measures of centrality based on betweenness. Sociometry **40**(1), 35–41 (1977)
20. Gallo, G., Longo, G., Pallottino, S., Nguyen, S.: Directed hypergraphs and applications. Discrete Appl. Math. **42**(2), 177–201 (1993)
21. George, B., Kim, S.: Spatio-Temporal Networks. Springer, New York (2003). doi:10.1007/978-1-4614-4918-8
22. George, B., Shekhar, S.: Time-aggregated graphs for modeling spatio-temporal networks. In: Advances in Conceptual Modeling-Theory and Practice, pp. 85–99. Springer, Berlin/Heidelberg (2006)
23. George, B., Kim, S., Shekhar, S.: Spatio-temporal network databases and routing algorithms: a summary of results. In: Proceedings of the 10th International Conference on Advances in Spatial and Temporal Databases, SSTD'07, pp. 460–477. Springer, Berlin/Heidelberg (2007)
24. George, B., Shekhar, S., Kim, S.: Spatio-temporal network databases and routing algorithms. Tech. Rep. 08-039, University of Minnesota - Computer Science and Engineering (2008)
25. Gunturi, V.M.V., Shekhar, S.: Lagrangian xgraphs: a logical data-model for spatio-temporal network data: a summary. In: Advances in Conceptual Modeling, pp. 201–211. Springer, Cham (2014)
26. Gunturi, V.M.V., Shekhar, S., Bhattacharya, A.: Minimum spanning tree on spatio-temporal networks. In: Proceedings of the 21st International Conference on Database and Expert Systems Applications: Part II, DEXA'10, pp. 149–158 (2010)
27. Gunturi, V.M.V., Nunes, E., Yang, K., Shekhar, S.: A critical-time-point approach to all-start-time lagrangian shortest paths: a summary of results. In: Advances in Spatial and Temporal Databases. Lecture Notes in Computer Science, vol. 6849, pp. 74–91. Springer, Berlin/Heidelberg (2011)
28. Gunturi, V.M.V., Shekhar, S., Joseph, K., Carley, K.M.: Scalable computational techniques for centrality metrics on temporally detailed social network. Mach. Learn. (2016). doi:10.1007/s10994-016-5583-7
29. Gunturi, V.M.V., Shekhar, S., Yang, K.: A critical-time-point approach to all-departure-time lagrangian shortest paths. IEEE Trans. Knowl. Data Eng. **27**(10), 2591–2603 (2015)
30. Habiba, Tantipathananandh, C., Berger-Wolf, T.Y.: Betweenness centrality measure in dynamic networks. Tech. Rep. 2007-19, Center for Discrete Mathematics and Theoretical Computer Science (2007)
31. Howison, J., Wiggins, A., Crowston, K.: Validity issues in the use of social network analysis with digital trace data. J. Assoc. Inf. Syst. **12**(12), Article 2 (2011)
32. Jing, N., Huang, Y.W., Rundensteiner, E.A.: Hierarchical optimization of optimal path finding for transportation applications. In: Proceedings of the Fifth International Conference on Information and Knowledge Management (CIKM), pp. 261–268. ACM, Rockville (1996)
33. Kamath, K.Y., Caverlee, J.: Transient crowd discovery on the real-time social web. In: Proceedings of ICWSM, pp. 585–594. ACM, New York (2011)
34. Kanoulas, E., Du, Y., Xia, T., Zhang, D.: Finding fastest paths on a road network with speed patterns. In: Proceedings of the 22nd International Conference on Data Engineering (ICDE), p. 10. IEEE, Atlanta (2006)

35. Karagiannis, T., Vojnovic, M.: Email information flow in large-scale enterprises. Technical Report Microsoft Research (2008)
36. Kargupta, H., Puttagunta, V., Klein, M., Sarkar, K.: On-board vehicle data stream monitoring using minefleet and fast resource constrained monitoring of correlation matrices. N. Gener. Comput. **25**(1), 5–32 (2006)
37. Kargupta, H., Gama, J., Fan, W.: The next generation of transportation systems, greenhouse emissions, and data mining. In: Proceedings of the 16th ACM SIGKDD International Conference on Knowledge Discovery and Data Mining, pp. 1209–1212. ACM, Washington, DC (2010)
38. Katz, L.: A new status index derived from sociometric analysis. Psychometrika **18**(1), 39–43 (1953)
39. Kaufman, D.E., Smith, R.L.: Fastest paths in time-dependent networks for intelligent vehicle-highway systems application. I V H S J. **1**(1), 1–11 (1993)
40. Kim, H., Anderson, R.: Temporal node centrality in complex networks. Phys. Rev. E **85**(2), 026107 (2012)
41. Kim, H., Tang, J., Anderson, R., Mascolo, C.: Centrality prediction in dynamic human contact networks. Comput. Netw. **56**(3), 983–996 (2012)
42. Kleinberg, J., Tardos, E.: Algorithm Design. Pearson Education, London (2009)
43. Köhler, E., Langkau, K., Skutella, M.: Time-expanded graphs for flow-dependent transit times. In: Proceedings of the 10th Annual European Symposium on Algorithms, ESA '02, pp. 599–611. Springer, London (2002)
44. Kossinets1, G., Watts, D.J.: Empirical analysis of an evolving social network. Science **311**(5757), 88–90 (2006)
45. Kulldorff, M.: A spatial scan statistic. Commun. Stat. Theory Methods **26**(6), 1481–1496 (1997)
46. Lehmann, J., et al.: Transient news crowds in social media. In: ICWSM (2013)
47. Lerman, K., Ghosh, R., Kang, J.H.: Centrality metric for dynamic network analysis. In: Proceedings of KDD Workshop on Mining and Learning with Graphs (MLG) (2010)
48. Liu, H., Hu, H.: Smart-signal phase ii: arterial offset optimization using archived high-resolution traffic signal data. Tech. Rep. CTS 13–19, Intel. Trans. Sys. Inst., Center for Transportation Studies, Univ. of Minnesota (Apr-2013)
49. Malmgren, R.D., et al.: A poissonian explanation for heavy tails in e-mail communication. Proc. Natl. Acad. Sci. **105**(47), 18153–18158 (2008)
50. Malmgren, R.D., et al.: Characterizing individual communication patterns. In: Proceedings of the 15th ACM SIGKDD International Conference on Knowledge Discovery and Data Mining, KDD '09, pp. 607–616 (2009)
51. Nannicini, G.: Point-to-point shortest paths on dynamic time-dependent road networks. 4OR **8**(3), 327–330 (2010)
52. Nannicini, G., Delling, D., Liberti, L., Schultes, D.: Bidirectional a* search for time-dependent fast paths. In: Experimental Algorithms, pp. 334–346. Springer, Berlin/Heidelberg (2008)
53. Nannicini, G., Delling, D., Schultes, D., Liberti, L.: Bidirectional a* search on time-dependent road networks. Networks **59**(2), 240–251 (2012)
54. NAVTEQ: Retrieved Oct 2017, https://www.here.com/en/navteq
55. Neill, D.B., Moore, A.W.: Rapid detection of significant spatial clusters. In: Proceedings of the Tenth ACM SIGKDD International Conference on Knowledge Discovery and Data Mining, KDD '04, pp. 256–265. ACM, Seattle (2004)
56. Nia, R., Bird, C., Devanbu, P., Filkov, V.: Validity of network analyses in open source projects. In: 2010 7th IEEE Working Conference on Mining Software Repositories (MSR), pp. 201–209 (2010)
57. Orda, A., Rom, R.: Shortest-path and minimum-delay algorithms in networks with time-dependent edge-length. J. ACM **37**(3), 607–625 (1990)
58. Panzarasa, P., et al.: Patterns and dynamics of users' behavior and interaction: Network analysis of an online community. J. Am. Soc. Inf. Sci. Technol. **60**(5), 911–932 (2009)

59. Shekhar, S., Liu, D.: Ccam: a connectivity-clustered access method for networks and network computations. IEEE Trans. Knowl. Data Eng. **9**(1), 102–119 (1997)

60. Shekhar, S., Gunturi, V., Evans, M.R., Yang, K.: Spatial big-data challenges intersecting mobility and cloud computing. In: MobiDE, pp. 1–6. ACM, New York (2012)

61. Tang, J., Musolesi, M., Mascolo, C., Latora, V., Nicosia, V.: Analysing information flows and key mediators through temporal centrality metrics. In: Proceedings of the 3rd Workshop on Social Network Systems, SNS '10, pp. 3:1–3:6 (2010)

62. Tantipathananandh, C., Berger-Wolf, T.Y.: Finding communities in dynamic social networks. In: IEEE 11th International Conference on Data Mining (ICDM), pp. 1236–1241 (2011)

63. Tantipathananandh, C., Berger-Wolf, T.: Constant-factor approximation algorithms for identifying dynamic communities. In: Proceedings of the 15th ACM SIGKDD International Conference on Knowledge Discovery and Data Mining, pp. 827–836. ACM, Paris (2009)

64. Tantipathananandh, C., Berger-Wolf, T., Kempe, D.: A framework for community identification in dynamic social networks. In: Proceedings of the 13th ACM SIGKDD International Conference on Knowledge Discovery and Data Mining, KDD '07, pp. 717–726. ACM, New York, NY (2007)

65. Tyler, J.R., Tang, J.C.: When can i expect an email response? A study of rhythms in email usage. In: Proceedings of the Eighth Conference on European Conference on Computer Supported Cooperative Work, pp. 239–258 (2003)

66. Wu, F., Huberman, B.A.: Novelty and collective attention. Proc. Natl. Acad. Sci. USA **104**(45), 17599–17601 (2007)

67. Xie, J., Kelley, S., Szymanski, B.K.: Overlapping community detection in networks: the state-of-the-art and comparative study. ACM Comput. Surv. **45**(4), 43:1–43:35 (2013)

68. Yuan, J., Zheng, Y., Zhang, C., Xie, W., Xie, X., Sun, G., Huang, Y.: T-drive: driving directions based on taxi trajectories. In: Proceedings of the SIGSPATIAL International Conference on Advances in GIS, GIS '10, pp. 99–108 (2010)

69. Zheng, Y., Zhou, X.E. (eds.): Computing with Spatial Trajectories. Springer, New York (2011)

Printed in the United States
By Bookmasters